CONTENTS

D0308202

AUTHORS' NOTE

This volume covers all the objectives included in the TEC Mathematics unit U80/713, comprising the two half-units U80/689 (Calculus) and U80/690 (Algebra). It is intended for students of all disciplines using the Mathematics unit.

Volumes covering the new Level I and Level II units have already been published. The three volumes now provide a full course of study for all the Mathematics required for the TEC Certificate.

A Greer
G W Taylor

1. DIFFERENTIATION

On reaching the end of this chapter you should be able to:

1. Use the derivatives of the functions: ax^n, $\sin ax$, $\cos ax$, $\tan x$, $\log_e x$, and e^{ax}.
2. Define the differential property of the exponential function.
3. Calculate the derivative at a point of the functions in 1.
4. State the basic rules of differential calculus for the derivatives of sum, product, quotient, and function of a function.
5. Determine the derivatives of various combinations of any two of the functions in 1 using 4.
6. Evaluate the derivatives in 5 at a given point.

INTRODUCTION

You may recall that the process of differentiation is a method of finding the rate of change of a function. The rate of change at a point on a curve may be found by determining the gradient of the tangent at that point. This may be achieved by either a theoretical or graphical method as follows.

Suppose we wish to find the gradient of the curve $y = x^2$ at the points where $x = 3$ and $x = -2$.

From a theoretical approach you may remember that if the function is of the form $y = x^n$, then the process of differentiating gives $\dfrac{dy}{dx} = nx^{n-1}$.

Hence if
$$y = x^2$$

then
$$\frac{dy}{dx} = 2x$$

and therefore when $x = 3$, the gradient of the tangent

$$\frac{dy}{dx} = 2(3) = +6$$

and when $x = -2$, the gradient of the tangent

$$\frac{dy}{dx} = 2(-2) = -4$$

1

Alternatively if we decide to use the graphical method we must first draw the graph as shown in Fig. 1.1.

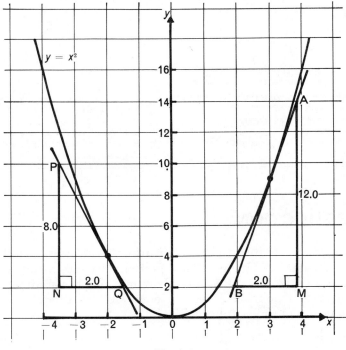

Fig. 1.1

Tangents have been drawn at the given points, that is where $x = 3$ and $x = -2$, and then the gradients may be found by constructing suitable right-angled triangles.

Where $\quad x = 3 \quad$ the gradient $= \dfrac{AM}{BM} = \dfrac{12.0}{2.0} = +6$

and where $\quad x = -2 \quad$ the gradient $= \dfrac{PN}{QN} = -\dfrac{8.0}{2.0} = -4$

These results verify those obtained by the theoretical method.

DIFFERENTIATION OF A SUM

To differentiate an expression containing the sum of several terms, we differentiate each individual term separately.

Hence if
$$y = x^4 + 2x^3 + 5x^2 + 7$$

then
$$\frac{dy}{dx} = 4x^3 + 6x^2 + 10x$$

And if
$$y = ax^3 + bx^2 - cx + d$$

then
$$\frac{dy}{dx} = 3ax^2 + 2bx - c$$

And if
$$y = \sqrt{x} + \frac{1}{\sqrt{x}} = x^{1/2} + x^{-1/2}$$

then
$$\frac{dy}{dx} = \tfrac{1}{2}x^{-1/2} + (-\tfrac{1}{2})x^{-3/2} = \frac{1}{2\sqrt{x}} - \frac{1}{2\sqrt{x^3}}$$

And if
$$y = 3.1x^{1.4} - \frac{3}{x} + 5 = 3.1x^{1.4} - 3x^{-1} + 5$$

then
$$\frac{dy}{dx} = (3.1)(1.4)x^{0.4} - 3(-1)x^{-2}$$

$$= 4.34x^{0.4} + \frac{3}{x^2}$$

And if
$$y = \frac{t^3 + t}{t^2} = \frac{t^3}{t^2} + \frac{t}{t^2} = t + t^{-1}$$

then
$$\frac{dy}{dt} = 1 + (-1)t^{-2} = 1 - \frac{1}{t^2}$$

Exercise 1.1

1) Find $\frac{dy}{dx}$ if $y = 5x^3 + 7x^2 - x - 1$.

2) If $s = 7\sqrt{t} - 6t^{0.3}$, find an expression for $\frac{ds}{dt}$.

3) Find by a theoretical method the value of $\frac{dy}{dx}$ when $x = 2$ for the curve $y = x - \frac{1}{x}$.

4) Find an expression for $\dfrac{dy}{du}$ if $y = \dfrac{u + u^2}{u}$.

5) Find graphically the gradient of the curve $y = x^2 + x + 2$ at the points where $x = +2$ and $x = -2$. Check the result by differentiation.

FUNCTIONS

If two variables x and y are connected so that the value of y depends upon the value allocated to x, then y is said to be a *function* of x. Thus if $y = 3x^2 + 4x - 7$, when $x = 5$, $y = 3 \times 5^2 + 4 \times 5 - 7 = 88$. Since the value of y depends upon the value allocated to x, y then is a function of x.

DIFFERENTIATION USING 'FUNCTION OF A FUNCTION'

Up to now we have only learnt how to differentiate comparatively simple expressions such as $y = 3x^{4.5}$. For more difficult expressions such as $y = \sqrt{(x^3 + 3x - 9)}$ we make a substitution.

If we put $u = x^3 + 3x - 9$ then $y = \sqrt{u}$.

Now y is a function of u, and since u is a function of x, it follows that y is a function of a function of x.

This all sounds rather complicated and the words 'function of a function' should be noted in case you meet them again, but we prefer to use the expression 'differentiation by substitution'.

DIFFERENTIATION BY SUBSTITUTION

A substitution method is often used for differentiating the more complicated expressions, together with the formula

$$\boxed{\dfrac{dy}{dx} = \dfrac{dy}{du} \times \dfrac{du}{dx}}$$

EXAMPLE 1.1

Find $\dfrac{dy}{dx}$ if $y = (x^2 - x)^9$.

We have $\qquad y = (x^2 - x)^9$

Then $\qquad y = u^9 \qquad$ where $\quad u = x^2 - x$

$\therefore \qquad \dfrac{dy}{du} = 9u^8 \qquad$ and $\qquad \dfrac{du}{dx} = 2x - 1$

But $\qquad \dfrac{dy}{dx} = \dfrac{dy}{du} \times \dfrac{du}{dx}$

$\therefore \qquad \dfrac{dy}{dx} = 9u^8 \times (2x - 1)$

The differentiation has now been completed and it only remains to put u in terms of x by using our original substitution $u = x^2 - x$.

Hence $\qquad \dfrac{dy}{dx} = 9(x^2 - x)^8 (2x - 1)$

EXAMPLE 1.2

Find $\dfrac{d}{dx}\left(\sqrt{(1 - 5x^3)}\right)$.

$\dfrac{d}{dx}\left(\sqrt{(1 - 5x^3)}\right)$ is called the differential coefficient of $\sqrt{(1 - 5x^3)}$ with respect to x. This simply means that we have to differentiate the expression with respect to x. If we let $y = \sqrt{(1 - 5x^3)}$, then the problem is to find $\dfrac{dy}{dx}$.

Let $\qquad y = \sqrt{(1 - 5x^3)}$

i.e. $\qquad y = (1 - 5x^3)^{1/2}$

Then $\qquad y = u^{1/2} \qquad$ where $\quad u = 1 - 5x^3$

$\therefore \qquad \dfrac{dy}{du} = \tfrac{1}{2}u^{-1/2} \qquad$ and $\qquad \dfrac{du}{dx} = -15x^2$

But $\qquad \dfrac{dy}{dx} = \dfrac{dy}{du} \times \dfrac{du}{dx}$

$$\therefore \qquad \frac{dy}{dx} = \tfrac{1}{2}u^{-1/2} \times (-15x^2)$$

$$\therefore \qquad \frac{dy}{dx} = \tfrac{1}{2}(1-5x^3)^{-1/2}(-15x^2)$$

$$\therefore \qquad = -\frac{15}{2}x^2(1-5x^3)^{-1/2}$$

$$= -\frac{15x^2}{2\sqrt{1-5x^3)}}$$

Hence $\qquad \dfrac{d}{dx}\left(\sqrt{(1-5x^3)}\right) = -\dfrac{15x^2}{2\sqrt{(1-5x^3)}}$

DIFFERENTIATION OF A FUNCTION OF A FUNCTION BY RECOGNITION

Consider $y = (\quad)^n$ where any function of x can be written inside the bracket. Then differentiating with respect to x, we have

$$\frac{dy}{dx} = \frac{dy}{d(\quad)} \times \frac{d(\quad)}{dx}$$

Thus to differentiate an expression of the type $(\quad)^n$, first differentiate the bracket, treating it as a term similar to x^n. Then differentiate the function x inside the bracket. Finally, to obtain an expression for $\dfrac{dy}{dx}$, multiply these two results together.

EXAMPLE 1.3

Find $\dfrac{dy}{dx}$ if $y = (x^2 - 5x + 3)^5$.

Differentiating the bracket as a whole we have

$$\frac{dy}{d(\quad)} = 5(x^2 - 5x + 3)^4 \qquad [1]$$

Also the function inside the bracket is $x^2 - 5x + 3$.

Differentiating this gives

$$\frac{d(\ \ )}{dx} = 2x - 5 \qquad [2]$$

Thus multiplying the results [1] and [2] together gives

$$\frac{dy}{dx} = 5(x^2 - 5x + 3)^4 \times (2x - 5)$$

$$= 5(2x - 5)(x^2 - 5x + 3)^4$$

Hence by recognising the method we can differentiate directly.

If for example $y = (x^2 - 3x)^7$.

then $\qquad \dfrac{dy}{dx} = 7(x^2 - 3x)^6 \times (2x - 3)$

$$= 7(2x - 3)(x^2 - 3x)^6$$

TO FIND THE RATE OF CHANGE OF sin θ, i.e. $\dfrac{d}{d\theta}$ (sin θ)

The rate of change of a curve at any point is the gradient of the tangent at that point. We shall, therefore, find the gradient at various points on the graph of $\sin \theta$ and then plot the values of these gradients to obtain a new graph.

It is suggested that the reader follows the method given, plotting his own curves on graph paper.

First, we plot the graph of $y = \sin \theta$ from $\theta = 0°$ to $\theta = 90°$ using values of $\sin \theta$ which may be obtained from your calculator. The curve is shown in Fig. 1.2.

Consider point P on the curve, where $\theta = 45°$, and draw the tangent APM.

We can find the gradient of the tangent by constructing a suitable right-angled triangle AMN (which should be as large as conveniently possible for accuracy) and finding the value of $\dfrac{MN}{AN}$.

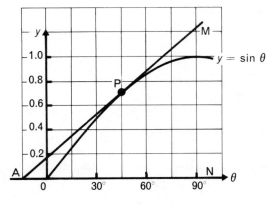

Fig. 1.2

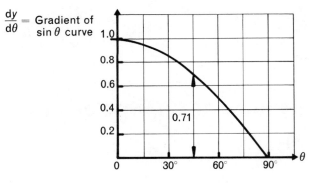

Fig. 1.3

Using the scale on the y-axis gives $MN = 1.29$ by measurement, and using the scale on the θ-axis gives $AN = 104°$ by measurement.

In calculations of this type it is necessary to obtain AN in radians. Remembering that

$$360° = 2\pi \text{ radians}$$

gives

$$1° = \frac{2\pi}{360} \text{ radians}$$

∴

$$104° = \frac{2\pi}{360} \times 104 = 1.81 \text{ radians}$$

Hence   Gradient at P $= \dfrac{MN}{AN} = \dfrac{1.29}{1.81} = 0.71$

The value 0.71 is used as the y-value at $\theta = 45°$ to plot a point on a new graph using the same scales as before. This new graph could be plotted on the same axes as $y = \sin\theta$ but for clarity it has been shown on new axes in Fig. 1.3.

This procedure is repeated for points on the $\sin\theta$ curve at θ values of $0°$, $15°$, $30°$, $60°$, $75°$ and $90°$, and the new curve obtained will be as shown in Fig. 1.3. This is the graph of the gradients of the sine curve at various points.

If we now plot a graph of $\cos\theta$, taking values from tables, on the axes in Fig. 1.3, we shall find that the two curves coincide — any difference will be due to errors from drawing the tangents.

Hence the gradient of the $\sin\theta$ curve at any value of θ is the same as the value of $\cos\theta$.

In other words, the rate of change of $\sin\theta$ is $\cos\theta$, provided that the angle θ is in radians.

In the above work we have only considered the graphs between $0°$ and $90°$ but the results are true for all values of the angle.

Hence if $\qquad y = \sin\theta \quad$ then $\quad \dfrac{dy}{dx} = \cos\theta$

or $\qquad \boxed{\dfrac{d}{d\theta}(\sin\theta) = \cos\theta} \qquad$ provided that θ is in radians.

The same procedure may be used to show that

$$\boxed{\dfrac{d}{d\theta}(\cos\theta) = -\sin\theta} \qquad \text{provided that } \theta \text{ is in radians.}$$

EXAMPLE 1.4

Find $\dfrac{d}{d\theta}(\sin 7\theta)$.

Let $\qquad\qquad y = \sin 7\theta$

Then $\qquad\qquad y = \sin u \qquad\qquad$ where $\quad u = 7\theta$

$\therefore \qquad\qquad \dfrac{dy}{du} = \cos u \qquad\qquad$ and $\quad \dfrac{du}{d\theta} = 7$

But $$\frac{dy}{d\theta} = \frac{dy}{du} \times \frac{du}{d\theta}$$

$\therefore$ $$\frac{dy}{d\theta} = (\cos u) \times 7 = 7 \cos u = 7 \cos 7\theta$$

$\therefore$ $$\frac{d}{d\theta}(\sin 7\theta) = 7 \cos 7\theta$$

EXAMPLE 1.5

Find $\dfrac{d}{d\theta}(\cos 4\theta)$.

Let $$y = \cos 4\theta$$

Then $$y = \cos u \qquad \text{where} \quad u = 4\theta$$

$\therefore$ $$\frac{dy}{du} = -\sin u \qquad \text{and} \quad \frac{du}{d\theta} = 4$$

But $$\frac{dy}{d\theta} = \frac{dy}{du} \times \frac{du}{d\theta}$$

$\therefore$ $$\frac{dy}{d\theta} = (-\sin u) \times 4 = -4 \sin u = -4 \sin 4\theta$$

$\therefore$ $$\frac{d}{d\theta}(\cos 4\theta) = -4 \sin 4\theta$$

In general
$$\frac{d}{dx}(\sin ax) = a \cos ax$$

and
$$\frac{d}{dx}(\cos ax) = -a \sin ax$$

EXAMPLE 1.6

Find $\dfrac{d}{dt}\left\{\cos\left(2t - \dfrac{3\pi}{2}\right)\right\}$.

Let $$y = \cos\left(2t - \frac{3\pi}{2}\right)$$

Then $\qquad y = \cos u \qquad$ where $\quad u = 2t - \dfrac{3\pi}{2}$

$\therefore \qquad \dfrac{dy}{du} = -\sin u \qquad$ and $\quad \dfrac{du}{dt} = 2$

But $\qquad \dfrac{dy}{dt} = \dfrac{dy}{du} \times \dfrac{du}{dt}$

$\therefore \qquad \dfrac{dy}{dt} = (-\sin u) \times 2 = -2\sin u$

$$= -2\sin\left(2t - \dfrac{3\pi}{2}\right)$$

$\therefore \qquad \dfrac{d}{dt}\left\{\cos\left(2t - \dfrac{3\pi}{2}\right)\right\} = -2\sin\left(2t - \dfrac{3\pi}{2}\right)$

THE DIFFERENTIAL COEFFICIENT OF $\log_e x$, i.e. $\dfrac{d}{dx}(\log_e x)$

In higher mathematics all logarithms are taken to the base e, where e = 2.718 28. Logarithms to this base are often called natural logarithms. They are also called Napierian or hyperbolic logarithms and are given as $\log_e$ or ln.

Again the graphical method of differentiation may be used. Fig. 1.4 shows the graph of $\log_e x$.

The reader may find it instructive to plot the curve of $y = \log_e x$ as shown in Fig. 1.4 and follow the procedure, as used previously, of drawing tangents at various points. The values of their gradients are then plotted, and a curve will result as shown in Fig. 1.5.

If a graph of $\dfrac{1}{x}$ is plotted on the same axes as in Fig. 1.5, it will be found to coincide with the gradient curve, that is, the $\dfrac{dy}{dx}$ graph.

Hence $\qquad \boxed{\dfrac{d}{dx}(\log_e x) = \dfrac{1}{x}}$

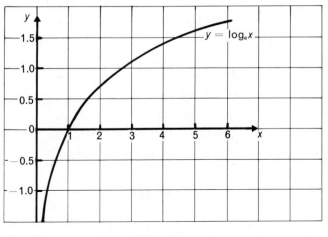

Fig. 1.4

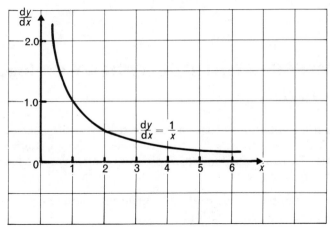

Fig. 1.5

EXAMPLE 1.7

Find $\dfrac{d}{dx}\{\log_e(x^2 + 5)\}$

Let $y = \log_e(x^2 + 5)$

Then $y = \log_e u$ where $u = x^2 + 5$

$\therefore$ $\dfrac{dy}{du} = \dfrac{1}{u}$ and $\dfrac{du}{dx} = 2x$

But
$$\frac{dy}{dx} = \frac{dy}{du} \times \frac{du}{dx}$$

$\therefore$
$$\frac{dy}{dx} = \frac{1}{u} \times 2x = \frac{1}{x^2 + 5} \times 2x$$

Hence
$$\frac{d}{dx}\{\log_e(x^2 + 5)\} = \frac{2x}{x^2 + 5}$$

THE DIFFERENTIAL COEFFICIENT OF THE EXPONENTIAL FUNCTION e^x, i.e. $\dfrac{\mathbf{d}}{\mathbf{dx}}(e^x)$

If we let $y = e^x$ then we need to find $\dfrac{dy}{dx}$.

Then
$$e^x = y$$

and rearranging in log form we get
$$x = \log_e y$$

$\therefore$
$$\frac{dx}{dy} = \frac{1}{y}$$

Hence by inverting both sides
$$\frac{dy}{dx} = y$$

But we know that $y = e^x$ and

$\therefore$
$$\frac{dy}{dx} = e^x$$

Hence
$$\frac{d}{dx}(e^x) = e^x$$

This result illustrates an important property of the exponential function, namely:

> The exponential function, e^x, has a differential coefficient of e^x which, therefore, is equal to the function itself.

EXAMPLE 1.8

Find $\dfrac{d}{dx}(e^{6x})$.

Let	$y = e^{6x}$	
Then	$y = e^u$	where $u = 6x$
$\therefore$	$\dfrac{dy}{du} = e^u$	and $\dfrac{du}{dx} = 6$
But	$\dfrac{dy}{dx} = \dfrac{dy}{du} \times \dfrac{du}{dx}$	
$\therefore$	$\dfrac{dy}{dx} = e^u \times 6 = e^{6x} \times 6 = 6e^{6x}$	

Hence

$$\frac{d}{dx}(e^{6x}) = 6e^{6x}$$

EXAMPLE 1.9

Find $\dfrac{d}{dx}\left(\dfrac{1}{e^{3x}}\right)$.

Let	$y = \dfrac{1}{e^{3x}}$	
or	$y = e^{-3x}$	
Then	$y = e^u$	where $u = -3x$
$\therefore$	$\dfrac{dy}{du} = e^u$	and $\dfrac{du}{dx} = -3$
But	$\dfrac{dy}{dx} = \dfrac{dy}{du} \times \dfrac{du}{dx}$	
$\therefore$	$\dfrac{dy}{dx} = e^u \times (-3) = -3e^u = -3e^{-3x}$	

$$\therefore \quad \frac{d}{dx}\left(\frac{1}{e^{3x}}\right) = \frac{d}{dx}(e^{-3x}) = -3e^{-3x}$$

In general

$$\boxed{\frac{d}{dx}(e^{ax}) = ae^{ax}}$$

We may now summarise differential coefficients of the more common functions:

y	$\dfrac{dy}{dx}$
ax^n	anx^{n-1}
$\sin ax$	$a \cos ax$
$\cos ax$	$-a \sin ax$
$\log_e x$	$\dfrac{1}{x}$
e^{ax}	ae^{ax}

At this stage we may use the results in the table to differentiate many functions by 'recognition' rather than using the method of substitution.

For example $$\frac{d}{d\theta}(\sin 3\theta) = 3 \cos 3\theta$$

or $$\frac{d}{dx}(e^{-3x}) = -3e^{-3x}$$

However, if the functions to be differentiated are not similar to those shown in the table, perhaps because they are more complicated, the method of substitution should be used.

Example 1.10 which follows shows how both the method of substitution and recognition may be used together to differentiate a more complicated function.

EXAMPLE 1.10

Find $\dfrac{dy}{dx}$ if $y = \sin^3 5x$.

We have $\qquad y = (\sin 5x)^3$

Then $\qquad y = u^3 \qquad\qquad$ where $\quad u = \sin 5x$

$\therefore \qquad \dfrac{dy}{du} = 3u^2 \qquad\qquad$ and $\quad \dfrac{du}{dx} = 5 \cos 5x$

But $\qquad \dfrac{dy}{dx} = \dfrac{dy}{du} \times \dfrac{du}{dx}$

$\therefore \qquad \dfrac{dy}{dx} = 3u^2 \times 5\cos 5x = 3\sin^2 5x \times 5\cos 5x$

$\therefore \qquad \dfrac{dy}{dx} = 15\sin^2 5x \cos 5x$

Exercise 1.2

Differentiate with respect to x:

1) $(3x+1)^2$

2) $(2-5x)^3$

3) $(1-4x)^{1/2}$

4) $(2-5x)^{3/2}$

5) $\dfrac{1}{4x^2+3}$

6) $\sin(3x+4)$

7) $\cos(2-5x)$

8) $\sin^2 4x$

9) $\dfrac{1}{\cos^3 7x}$

10) $\sin\left(2x+\dfrac{\pi}{2}\right)$

11) $\cos^3 x$

12) $\dfrac{1}{\sin x}$

13) $\log_e 9x$

14) $9\log_e\left(\dfrac{5}{x}\right)$

15) $\tfrac{1}{4}\log_e(2x-7)$

16) $\dfrac{1}{e^x}$

17) $2e^{3x+4}$

18) $\dfrac{1}{e^{2-8x}}$

19) Find $\dfrac{d}{dt}\left(\dfrac{1}{\sqrt[3]{1-2t}}\right)$

20) Find $\dfrac{d}{d\theta}\{\sin(\tfrac{3}{4}\theta - \pi)\}$

21) Find $\dfrac{d}{d\phi}\left(\dfrac{1}{\cos(\pi-\phi)}\right)$

22) Find $\dfrac{d}{dx}\left(\log_e\dfrac{1}{\sqrt{x}}\right)$

23) Find $\dfrac{d}{dt}(Be^{kt-b})$

24) Find $\dfrac{d}{dx}(\sqrt[3]{e^{1-x}})$

DIFFERENTIATION OF A PRODUCT

If $y = u \times v$, where u and v are functions of x, we must use the formula

$$\frac{dy}{dx} = v\frac{du}{dx} + u\frac{dv}{dx}$$

EXAMPLE 1.11

Find $\dfrac{d}{dx}(x^3 \sin 2x)$.

Let $\quad y = x^3 \sin 2x$

Then $\quad y = u \times v \quad$ where $\quad u = x^3 \quad$ and $\quad v = \sin 2x$

$$\therefore \quad \frac{du}{dx} = 3x^2 \text{ and } \frac{dv}{dx} = 2\cos 2x$$

But $\quad \dfrac{dy}{dx} = v\dfrac{du}{dx} + u\dfrac{dv}{dx}$

$$\therefore \quad \frac{dy}{dx} = (\sin 2x)3x^2 + x^3(2\cos 2x) = x^2(3\sin 2x + 2x\cos 2x)$$

Then $\quad \dfrac{d}{dx}(x^3 \sin 2x) = x^2(3\sin 2x + 2x\cos 2x)$

EXAMPLE 1.12

Differentiate $(x^2 + 1)\log_e x$ with respect to x.

If we let $y = (x^2 + 1)\log_e x$ then the problem is to find $\dfrac{dy}{dx}$.

Then $\quad y = u \times v \quad$ where $\quad u = x^2 + 1 \quad$ and $\quad v = \log_e x$

$$\therefore \quad \frac{du}{dx} = 2x \qquad \text{and } \frac{dv}{dx} = \frac{1}{x}$$

But $\quad \dfrac{dy}{dx} = v\dfrac{du}{dx} + u\dfrac{dv}{dx}$

$$\therefore \quad \frac{dy}{dx} = (\log_e x)2x + (x^2 + 1)\frac{1}{x} = 2x(\log_e x) + x + \frac{1}{x}$$

DIFFERENTIATION OF A QUOTIENT

If $y = \dfrac{u}{v}$, where u and v are functions of x, we must use the formula

$$\frac{dy}{dx} = \frac{v\dfrac{du}{dx} - u\dfrac{dv}{dx}}{v^2}$$

EXAMPLE 1.13

Find $\dfrac{dy}{dx}$ if $y = \dfrac{e^{2x}}{x+3}$.

We have $y = \dfrac{e^{2x}}{x+3}$

Let $y = \dfrac{u}{v}$ where $u = e^{2x}$ and $v = x + 3$

$$\therefore \quad \frac{du}{dx} = 2e^{2x} \text{ and } \frac{dv}{dx} = 1$$

But $\dfrac{dy}{dx} = \dfrac{v\dfrac{du}{dx} - u\dfrac{dv}{dx}}{v^2}$

$\therefore \quad \dfrac{dy}{dx} = \dfrac{(x+3)2e^{2x} - e^{2x} \times 1}{(x+3)^2}$

$\qquad\qquad = \dfrac{e^{2x}(2x + 6 - 1)}{(x+3)^2}$

$\qquad\qquad = \dfrac{(2x+5)e^{2x}}{(x+3)^2}$

EXAMPLE 1.14

Find $\dfrac{d}{d\theta}(\tan\theta)$.

Let $y = \tan\theta$

or $y = \dfrac{\sin\theta}{\cos\theta}$

Then $y = \dfrac{u}{v}$ where $u = \sin\theta$ and $v = \cos\theta$

$$\therefore \quad \frac{du}{d\theta} = \cos\theta \quad \text{and} \quad \frac{dv}{d\theta} = -\sin\theta$$

But $\dfrac{dy}{d\theta} = \dfrac{v\dfrac{du}{d\theta} - u\dfrac{dv}{d\theta}}{v^2}$

$$\therefore \quad \frac{dy}{d\theta} = \frac{(\cos\theta)(\cos\theta) - (\sin\theta)(-\sin\theta)}{\cos^2\theta}$$

$$= \frac{\cos^2\theta + \sin^2\theta}{\cos^2\theta}$$

Using the identity $\sin^2\theta + \cos^2\theta = 1$

then $\dfrac{dy}{d\theta} = \dfrac{1}{\cos^2\theta} = \sec^2\theta$

Hence $\boxed{\dfrac{d}{d\theta}(\tan\theta) = \sec^2\theta}$ provided that θ is in radians.

Exercise 1.3

1) Differentiate with respect to x

   (a) $x\sin x$ (b) $e^x\tan x$ (c) $x\log_e x$

2) Find $\dfrac{d}{dt}(\sin t\cos t)$ 3) Find $\dfrac{d}{d\theta}(\sin 2\theta\tan\theta)$

4) Find $\dfrac{d}{dm}(e^{4m}\cos 3m)$ 5) Find $\dfrac{d}{dx}(3x^2\log_e x)$

6) Find $\dfrac{d}{dt}\{6e^{3t}(t^2 - 1)\}$ 7) Find $\dfrac{d}{dz}\{(z - 3z^2)\log_e z\}$

8) Differentiate with respect to x

   (a) $\dfrac{x}{1-x}$ (b) $\dfrac{\log_e x}{x^2}$ (c) $\dfrac{e^x}{\sin 2x}$

9) Find $\dfrac{d}{dz}\left(\dfrac{z+2}{3-4z}\right)$

10) Find $\dfrac{d}{dt}\left(\dfrac{\cos 2t}{e^{2t}}\right)$

11) Find $\dfrac{d}{d\theta}(\cot\theta)$ $\left(Hint:\ \text{Use the identity } \cot\theta = \dfrac{\cos\theta}{\sin\theta}\right)$.

NUMERICAL VALUES OF DIFFERENTIAL COEFFICIENTS

EXAMPLE 1.15

Find the value of $\dfrac{dy}{dx}$ for the curve $y = \dfrac{1}{\sqrt{x}} - 3\log_e x$ at the point where $x = 2.3$

We have
$$y = x^{-1/2} - 3\log_e x$$

$$\therefore \qquad \frac{dy}{dx} = -\tfrac{1}{2}x^{-3/2} - \frac{3}{x}$$

It is often difficult to decide how much simplification of an expression will help in finding its numerical value when a particular value of x is substituted. In this case the expression may be rewritten as:

$$\frac{dy}{dx} = -\frac{1}{2(\sqrt{x})^3} - \frac{3}{x}$$

Hence when
$$x = 2.3$$

then
$$\frac{dy}{dx} = -\frac{1}{2(\sqrt{2.3})^3} - \frac{3}{2.3} = -1.45$$

It may well be argued that if a scientific calculator is available the value of x may just as well be substituted into the original expression for $\dfrac{dy}{dx}$ without any simplification.

This would give $\dfrac{dy}{dx} = -\tfrac{1}{2}(2.3)^{-1.5} - \dfrac{3}{2.3}$ which can be evaluated as easily as the 'simplified' arrangement.

It may, however, be more difficult to detect a computation error when making a rough check of the answer which is why expressions with positive indices are often preferred.

EXAMPLE 1.16

A curve is given in the form $y = 3 \sin 2\theta - 5 \tan \theta$ when θ is in radians. Find the gradient of the curve at the point where θ has a value equivalent to $34°$.

The gradient of the curve is given by $\dfrac{dy}{d\theta}$.

We have
$$y = 3 \sin 2\theta - 5 \tan \theta$$

$$\frac{dy}{d\theta} = 3 \times 2 \cos 2\theta - 5 \sec^2\theta$$

Substituting the value $\theta = 34°$

gives
$$\frac{dy}{d\theta} = 6 \cos (2 \times 34)° - 5 \sec^2 (34)°$$

$$= 6 \cos 68° - 5(\sec 34°)^2$$

$$= -5.03$$

EXAMPLE 1.17

If $y = \frac{1}{2}[e^{3t} + e^{-3t}]$ find the value of $\dfrac{dy}{dt}$ when $t = -0.63$

We have
$$y = \frac{1}{2}[e^{3t} + e^{-3t}]$$

$$\frac{dy}{dt} = \frac{1}{2}[3e^{3t} + (-3)e^{-3t}]$$

Substituting the value $t = -0.63$

gives
$$\frac{dy}{dt} = \frac{3}{2}[e^{3(-0.63)} - e^{-3(-0.63)}]$$

$$= \frac{3}{2}[e^{-1.89} - e^{1.89}]$$

$$= \frac{3}{2}[0.151 - 6.619]$$

$$= -9.70$$

EXAMPLE 1.18

Find the gradient of the curve $\dfrac{\cos x}{x}$ at the point where $x = 0.25$

The gradient of the curve is given by $\dfrac{dy}{dx}$ if we let

$$y = \frac{\cos x}{x}$$

Then $\qquad y = \dfrac{u}{v} \qquad$ where $\qquad u = \cos x \qquad$ and $\qquad v = x$

$$\therefore \qquad \frac{du}{dx} = -\sin x \qquad \therefore \qquad \frac{dv}{dx} = 1$$

But $\qquad \dfrac{dy}{dx} = \dfrac{v\dfrac{du}{dx} - u\dfrac{dy}{dx}}{v^2}$

$$\therefore \qquad \frac{dy}{dx} = \frac{x(-\sin x) - (\cos x)1}{x^2}$$

$$= \frac{-x \sin x - \cos x}{x^2}$$

If we now substitute the value $x = 0.25$

then $\qquad \dfrac{dy}{dx} = \dfrac{-(0.25)(\sin 0.25) - (\cos 0.25)}{(0.25)^2}$

The value 0.25 must be treated as radians when substituted into trigonometrical functions such as $\sin x$ and $\cos x$.

If a scientific calculator is used it is usually possible to set the machine to accept radians and give directly trigonometrical ratios.

Then $\qquad \dfrac{dy}{dx} = -16.5$

If tables are to be used to perform the calculation it may well be necessary to convert the angle to degrees.

Now

$$0.25 \, \text{rad} = \left(0.25 \times \frac{180}{\pi}\right)^{\circ} = 14°19'$$

and hence

$$\frac{dy}{dx} = \frac{-(0.25)\sin 14°19' - \cos 14°19'}{(0.25)^2}$$

$$= -16.5$$

Exercise 1.4

1) If $y = 3x^2 - \dfrac{7}{x^2} + \sqrt{x}$, find the value of $\dfrac{dy}{dx}$ if $x = 3.5$.

2) If $y = 5\sin 2\theta + 3\cos\dfrac{\theta}{2}$, find the value of $\dfrac{dy}{d\theta}$ if

$$\theta = 0.942 \text{ radians}$$

3) Find the value of $\dfrac{dy}{dt}$ when $t = -0.1$ if $y = \frac{1}{2}(e^t - e^{-t})$.

4) If $x = 0.3$, find the value of $\dfrac{dy}{dx}$ when $y = \sqrt{(3 - 2x^2)}$.

5) If $y = \sin^4 2\theta$, find the value of $\dfrac{dy}{d\theta}$ when $\theta = \dfrac{3\pi}{2}$ radians.

6) A curve is given in the form $y = \tan(3\phi - \pi)$, where ϕ is in radians. Find the gradient of the curve at the point where ϕ has a value equivalent to $23.4°$.

7) If $y = 4\log_e(1 - x)$, find the value of $\dfrac{dy}{dx}$ when $x = 0.32$

8) Find the value of $\dfrac{dy}{dx}$ when $x = 2.9$ when $y = e^{(9-3x)}$.

9) Given that $y = (\sin x)(\cos x)$ find the value of $\dfrac{dy}{dx}$ when

$$x = \dfrac{\pi}{6} \text{ radians}$$

10) If $y = \dfrac{1 + x^2}{x - 2}$, find the value of $\dfrac{dy}{dx}$ if $x = -1.25$

On reaching the end of this chapter you should be able to:

1. *State the notation for second derivatives as $\frac{d^2y}{dx^2}$ and similar form, e.g. $\frac{d^2x}{dt^2}$.*

2. *Determine a second derivative, by applying the basic rules of differential calculus, to the simplified result of a first differentiation.*

3. *Evaluate a second derivative determined in 2*

at a given point.

4. *State that $\frac{ds}{dt}$ and $\frac{d^2s}{dt^2}$ express velocity and acceleration.*

5. *Calculate the velocity and acceleration at a given time from an equation for displacement expressed in terms of time using 4.*

INTRODUCTION

If
$$y = x^6$$

then
$$\frac{dy}{dx} = 6x^5$$

and if we differentiate this equation again with respect to x we obtain

$$\frac{d}{dx}\left(\frac{dy}{dx}\right) = \frac{d}{dx}(6x^5)$$

or
$$\frac{d^2y}{dx^2} = 30x^4$$

Now just as $\frac{dy}{dx}$ is called the *first differential coefficient*, or *first derivative*, of y with respect to x, so $\frac{d^2y}{dx^2}$ is called the *second differential coefficient*, or *second derivative*, of y with respect to x.

It should be noted that the figure 2 which occurs twice in $\frac{d^2y}{dx^2}$ is *not* an index but merely indicates that the original function has been differentiated twice. Hence $\frac{d^2y}{dx^2}$ is *not* the same as $\left(\frac{dy}{dx}\right)^2$.

EXAMPLE 2.1

If $y = x^3 - 2x^2 + 3x - 7$ find $\dfrac{dy}{dx}$ and $\dfrac{d^2y}{dx^2}$.

Now
$$y = x^3 - 2x^2 + 3x - 7$$

$\therefore$
$$\frac{dy}{dx} = 3x^2 - 4x + 3$$

and
$$\frac{d^2y}{dx^2} = 6x - 4$$

EXAMPLE 2.2

If $y = \sqrt{x} + \log_e x$ find the values of $\dfrac{dy}{dx}$ and $\dfrac{d^2y}{dx^2}$ when $x = 2$.

Now
$$y = x^{1/2} + \log_e x$$

$\therefore$
$$\frac{dy}{dx} = \tfrac{1}{2}x^{-1/2} + \frac{1}{x}$$

$$= \tfrac{1}{2}x^{-1/2} + x^{-1}$$

$\therefore$
$$\frac{d^2y}{dx^2} = -\tfrac{1}{4}x^{-3/2} + (-1)x^{-2} = -\tfrac{1}{4}x^{-3/2} - x^{-2}$$

Before substituting the numerical value of x we recommend that the expressions for $\dfrac{dy}{dx}$ and $\dfrac{d^2y}{dx^2}$ are transformed to give positive indices.

Although most electronic calculators will evaluate expressions with negative indices, mistakes may occur and it is more difficult to spot a computation error.

Now
$$\frac{dy}{dx} = \frac{1}{2x^{1/2}} + \frac{1}{x} = \frac{1}{2\sqrt{x}} + \frac{1}{x}$$

Hence when $x = 2$

$$\frac{dy}{dx} = \frac{1}{2\sqrt{2}} + \frac{1}{2} = 0.354 + 0.5 = 0.854$$

Also
$$\frac{d^2y}{dx^2} = -\frac{1}{4x^{3/2}} - \frac{1}{x^2} = -\frac{1}{4(\sqrt{x})^3} - \frac{1}{x^2}$$

and when $x = 2$

$$\frac{d^2y}{dx^2} = -\frac{1}{4(\sqrt{2})^3} - \frac{1}{2^2} = -0.088 - 0.250 = -0.338$$

EXAMPLE 2.3

If $\theta = \dfrac{\pi}{2}$ find the values of $\dfrac{dy}{d\theta}$ and $\dfrac{d^2y}{d\theta^2}$ given that $y = \sin 2\theta + \cos 3\theta$.

Now
$$y = \sin 2\theta + \cos 3\theta$$

$\therefore$
$$\frac{dy}{d\theta} = 2\cos 2\theta - 3\sin 3\theta$$

and
$$\frac{d^2y}{d\theta^2} = -4\sin 2\theta - 9\cos 3\theta$$

If a value of an angle is given in terms of π the units are radians.

Therefore, when $\theta = \dfrac{\pi}{2}$ we have

$$\frac{dy}{d\theta} = 2\cos 2\left(\frac{\pi}{2}\right) - 3\sin 3\left(\frac{\pi}{2}\right)$$

$$= 2\cos 180° - 3\sin 270° = 2(-1) - 3(-1) = 1$$

and when $\theta = \dfrac{\pi}{2}$ we have

$$\frac{d^2y}{d\theta^2} = -4\sin 2\left(\frac{\pi}{2}\right) - 9\cos 3\left(\frac{\pi}{2}\right) = -4(0) - 9(0) = 0$$

EXAMPLE 2.4

If $y = \frac{1}{2}(e^{2t} + e^{-2t})$ find the values of $\dfrac{dy}{dt}$ and $\dfrac{d^2y}{dt^2}$ if $t = 0.61$

We have
$$y = \tfrac{1}{2}e^{2t} + \tfrac{1}{2}e^{-2t}$$

$\therefore$
$$\frac{dy}{dt} = \tfrac{1}{2}(2e^{2t}) + \tfrac{1}{2}(-2e^{-2t})$$

$$= e^{2t} - e^{-2t}$$

Also
$$\frac{d^2y}{dt^2} = 2e^{2t} - (-2)e^{-2t}$$

$$= 2e^{2t} + 2e^{-2t}$$

Hence when $t = 0.61$

$$\frac{dy}{dt} = e^{2(0.61)} - e^{-2(0.61)} = 3.09$$

and when $t = 0.61$

$$\frac{d^2y}{dt^2} = 2e^{2(0.61)} + 2e^{-2(0.61)} = 7.36$$

EXAMPLE 2.5

If $y = \tan\theta$ find the value of $\dfrac{d^2y}{d\theta^2}$ when $\theta = 0.436$ radians.

We have
$$y = \tan\theta$$

$\therefore$
$$\frac{dy}{d\theta} = \sec^2\theta$$

That is
$$\frac{dy}{d\theta} = (\cos\theta)^{-2}$$

$\therefore$
$$\frac{d^2y}{d\theta^2} = 2(\sin\theta)(\cos\theta)^{-3}$$

You are left to check this second differentiation by differentiating $(\cos\theta)^{-2}$ by the method of substitution.

Now when $\theta = 0.436$ radians

$$\frac{d^2y}{d\theta^2} = \frac{2(\sin 0.436)}{(\cos 0.436)^3} = 1.135$$

Exercise 2.1

1) If $y = 3x^3 + 2x - 7$, find an expression for $\dfrac{d^2y}{dx^2}$ and also its value when $x = 3$.

2) Find the values of $\dfrac{dy}{dx}$ and $\dfrac{d^2y}{dx^2}$ when $x = -2$ given that $y = 5x^4 + 7x^2 + x$.

3) Given that $y = \dfrac{3t^5 + 2t}{t^2}$ find the value of $\dfrac{d^2y}{dt^2}$ when $t = 0.6$

4) If $y = \log_e x$, find $\dfrac{d^2y}{dx^2}$ in terms of x, and also its value when $x = 1.9$

5) If $z = 5\cos 3\theta$, find the value of $\dfrac{dz}{d\theta}$ and $\dfrac{d^2z}{d\theta^2}$ when $\theta = \dfrac{\pi}{2}$ radians.

6) Find the value of $\dfrac{d^2p}{d\phi^2}$ given that $p = 6\sin 4\phi$, where ϕ is in radians, when ϕ has a value equivalent to $28°$.

7) If $y = 2\cos\dfrac{\alpha}{4}$, find the value of $\dfrac{d^2y}{d\alpha^2}$ when $\alpha = 1.6$ radians.

8) When $t = 2$, find the value of $\dfrac{d^2v}{dt^2}$ given that $v = e^t + e^{-t}$.

9) If $y = -e^{4.6t}$, find the value of $\dfrac{dy}{dt}$ and $\dfrac{d^2y}{dt^2}$ when $t = 0$.

10) Find the value of $\dfrac{d^2u}{dm^2}$ if $u = \tfrac{1}{2}(e^{3m} - e^{-3m})$ given that $m = 1.3$

11) If $y = x\log_e x$, find the value of $\dfrac{d^2y}{dx^2}$ if $x = 0.34$

VELOCITY AND ACCELERATION

Suppose that a vehicle starts from rest and travels 60 metres in 12 seconds. The average velocity may be found by dividing the total distance travelled by the total time taken, that is $\tfrac{60}{12} = 5\,\text{m/s}$. This is *not* the *instantaneous* velocity, however, *at* a time of 12 seconds, but is the *average velocity* over the 12 seconds as calculated previously.

Fig. 2.1 shows a graph of distance s, against time t. The average velocity over a period is given by the gradient of the chord which meets the curve at the extremes of the period. Thus in the diagram the gradient of the dotted chord QR gives the average velocity between $t = 2\,\text{s}$ and $t = 6\,\text{s}$. It is found to be $\tfrac{13}{4} = 3.25\,\text{m/s}$.

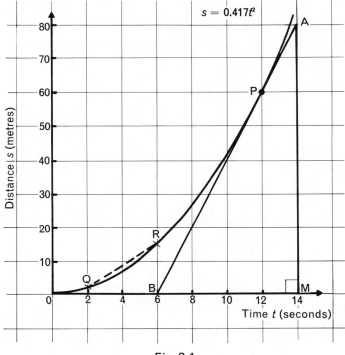

$$s = 0.417t^2$$

Fig. 2.1

The velocity at any point is the rate of change of s with respect t and may be found by finding the gradient of the curve at that point. In mathematical notation this is given by $\dfrac{ds}{dt}$.

Suppose we know that the relationship between s and t is

$$s = 0.417t^2$$

Then velocity, $\qquad v = \dfrac{ds}{dt} = 0.834t$

and hence when $t = 12$ seconds, $v = 0.834 \times 12 = 10$ m/s.

This result may be found graphically be drawing the tangent to the curve of s against t at the point P and constructing a suitable right-angled triangle ABM.

Hence the velocity at $P = \dfrac{AM}{BM} = \dfrac{80}{8} = 10$ m/s which verifies the theoretical result.

Similarly, the rate of change of velocity with respect to time is called acceleration and is given by the gradient of the velocity-time graph at any point. In mathematical notation this is given by $\dfrac{dv}{dt}$.

Now
$$\frac{dv}{dt} = \frac{d}{dt}(v) = \frac{d}{dt}\left(\frac{ds}{dt}\right) = \frac{d^2s}{dt^2}$$

and so the acceleration, a, is given by either

$$\frac{dv}{dt} \qquad \text{or} \qquad \frac{d^2s}{dt^2}$$

The above reasoning was applied to linear motion, but it could also have been used for angular motion. The essential difference is that distance, s, is replaced by angle turned through, θ rad.

Both sets of results are summarised in Fig. 2.2.

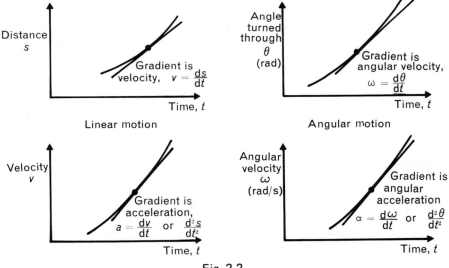

Fig. 2.2

EXAMPLE 2.6

A body moves a distance s metres in a time of t seconds so that $s = 2t^3 - 9t^2 + 12t + 6$. Find:

a) its velocity after 3 seconds,

b) its acceleration after 3 seconds, and

c) when the velocity is zero.

We have $$s = 2t^3 - 9t^2 + 12t + 6$$

$\therefore$ $$\frac{ds}{dt} = 6t^2 - 18t + 12$$

and $$\frac{d^2s}{dt^2} = 12t - 18$$

a) When $t = 3$, then the velocity is

$$\frac{ds}{dt} = 6(3)^2 - 18(3) + 12 = 12 \text{ m/s}$$

b) When $t = 3$, then the acceleration is $\dfrac{d^2s}{dt^2} = 12(3) - 18 = 18 \text{ m/s}^2$.

c) When the velocity is zero then $\dfrac{ds}{dt} = 0$.

That is $$6t^2 - 18t + 12 = 0$$

$\therefore$ $$t^2 - 3t + 2 = 0$$

$\therefore$ $$(t-1)(t-2) = 0$$

$\therefore$ either $$t - 1 = 0 \quad \text{or} \quad t - 2 = 0$$

$\therefore$ either $$t = 1 \text{ second or } t = 2 \text{ seconds}$$

EXAMPLE 2.7

The angle θ radians is connected with the time t seconds by the relationship $\theta = 20 + 5t^2 - t^3$. Find:

a) the angular velocity when $t = 2$ seconds, and

b) the value of t when the angular deceleration is 4 rad/s².

We have $$\theta = 20 + 5t^2 - t^3$$

$\therefore$ $$\frac{d\theta}{dt} = 10t - 3t^2$$

and $$\frac{d^2\theta}{dt^2} = 10 - 6t$$

a) When $t = 2$, then the angular velocity

$$\frac{d\theta}{dt} = 10(2) - 3(2)^2 = 8\,\text{rad/s}.$$

b) An angular deceleration of $4\,\text{rad/s}^2$ may be called an angular acceleration of $-4\,\text{rad/s}^2$.

$\therefore$ when $\qquad \dfrac{d^2\theta}{dt^2} = -4 \quad$ then $\quad -4 = 10 - 6t$

$$\text{or} \qquad t = 2.33\,\text{seconds}$$

Exercise 2.2

1) If $s = 10 + 50t - 2t^2$, where s metres is the distance travelled in t seconds by a body, what is the velocity of the body after 2 seconds?

2) If $v = 5 + 24t - 3t^2$ where v m/s is the velocity of a body at a time t seconds, what is the acceleration when $t = 3$?

3) A body moves s metres in t seconds where $s = t^3 - 3t^2 - 3t + 8$. Find:

(a) its velocity at the end of 3 seconds;

(b) when its velocity is zero;

(c) its acceleration at the end of 2 seconds;

(d) when its acceleration is zero.

4) A body moves s metres in t seconds where $s = \dfrac{1}{t^2}$. Find the velocity and acceleration after 3 seconds.

5) The distance s metres travelled by a falling body starting from rest after a time t seconds is given by $s = 5t^2$. Find its velocity after 1 second and after 3 seconds.

6) The distance s metres moved by the end of a lever after a time t seconds is given by the formula $s = 6t^2$. Find the velocity of the end of the lever when it has moved a distance $\frac{1}{2}$ metre.

7) The angular displacement θ radians of the spoke of a wheel is given by the expression $\theta = \frac{1}{2}t^4 - t^3$ where t seconds is the time. Find:

(a) the angular velocity after 2 seconds;

(b) the angular acceleration after 3 seconds;

(c) when the angular acceleration is zero.

8) An angular displacement θ radians in time t seconds is given by the equation $\theta = \sin 3t$. Find:

(a) the angular velocity when $t = 1$ second;

(b) the smallest positive value of t for which the angular velocity is 2 rad/s;

(c) the angular acceleration when $t = 0.5$ seconds;

(d) the smallest positive value of t for which the angular acceleration is 9 rad/s^2.

9) A mass of 5000 kg moves along a straight line so that the distance s metres travelled in a time t seconds is given by $s = 3t^2 + 2t + 3$. If v m/s is its velocity and m kg is its mass, then its kinetic energy is given by the formula $\frac{1}{2}mv^2$. Find its kinetic energy at a time $t = 0.5$ seconds remembering that the joule (J) is the unit of energy.

3. MAXIMUM AND MINIMUM

TURNING POINTS

At the points P and Q (Fig. 3.1) the tangent to the curve is parallel to the *x*-axis. The points P and Q are called *turning points*. The turning point at P is called a *maximum* turning point and the turning point at Q is called a *minimum* turning point. It will be seen from Fig. 3.1 that the value of *y* at P is not the greatest value of *y* nor is the value of *y* at Q the least. The terms maximum and minimum values apply only in the vicinity of the turning points and not to the values of *y* in general.

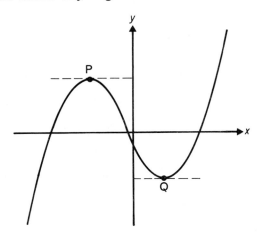

Fig. 3.1

In practical applications, however, we are usually concerned with a specific range of values of x which are dictated by the problem. There is then no difficulty in identifying a particular maximum or minimum within this range of values of x.

EXAMPLE 3.1

Plot the graph of $y = x^3 - 5x^2 + 2x + 8$ for values of x between -2 and 6. Hence find the maximum and minimum values of y.

To plot the graph we draw up a table in the usual way:

x	-2	-1	0	1	2	3	4	5	6
$y = x^3 - 5x^2 + 2x + 8$	-24	0	8	6	0	-4	0	18	56

The graph is shown in Fig. 3.2. The maximum value occurs at the point P where the tangent to the curve is parallel to the x-axis. The minimum values occurs at the point Q where again the tangent to the curve is parallel to the x-axis. From the graph the maximum value of y is 8.21 and the minimum value of y is -4.06

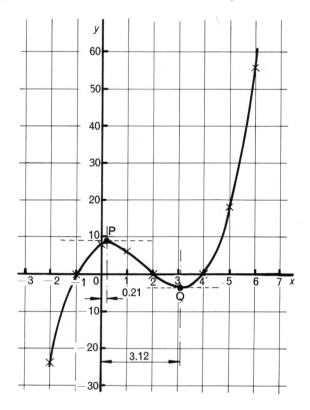

Fig. 3.2

Notice that the value of y at P is not the greatest value of y nor is the value of y at Q the least. However, the values of y at P and Q are called the 'maximum' and 'minimum' values of y respectively.

It is not always convenient to draw the full graph to find the turning points as in the previous example. At a turning point the tangent to the curve is parallel to the x-axis (Fig. 3.3) and hence the gradient of the curve is zero, i.e. $\dfrac{dy}{dx} = 0$. Using this fact enables us to find the values of x at which the turning points occur.

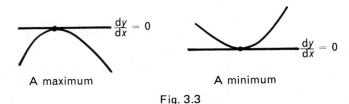

A maximum   A minimum

Fig. 3.3

MAXIMUM OR MINIMUM?

It is often necessary to determine whether each point is a maximum or a minimum.

Two methods of testing are as follows.

Method 1

Consider the gradients of the curve on either side of the turning point. Fig. 3.4 shows how the gradient (or slope) of curve changes in the vicinity of a turning point.

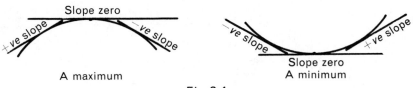

A maximum   A minimum

Fig. 3.4

Method 2

Find the value of $\dfrac{d^2y}{dx^2}$ at the turning point.

If it is positive then the turning point is a minimum, and if it is negative, then the turning point is a maximum.

If the original expression can be differentiated twice and the expression for $\dfrac{d^2y}{dx^2}$ obtained without too much difficulty, then the second method is generally used.

EXAMPLE 3.2

Find the maximum and minimum values of y given that
$$y = x^3 + 3x^2 - 9x + 6$$

We have $\qquad\qquad y = x^3 + 3x^2 - 9x + 6$

$\therefore \qquad\qquad\qquad \dfrac{dy}{dx} = 3x^2 + 6x - 9$

and $\qquad\qquad\qquad \dfrac{d^2y}{dx^2} = 6x + 6$

At a turning point $\qquad \dfrac{dy}{dx} = 0$

$\therefore \qquad\qquad 3x^2 + 6x - 9 = 0$

$\therefore \qquad\qquad x^2 + 2x - 3 = 0 \qquad$ by dividing through by 3

$\therefore \qquad\qquad (x - 1)(x + 3) = 0$

$\therefore$ either $\qquad\qquad x - 1 = 0 \quad$ or $\quad x + 3 = 0$

$\therefore$ either $\qquad\qquad\quad x = 1 \quad$ or $\qquad x = -3$

Test for maximum or minimum:

From above we have

$$\dfrac{d^2y}{dx^2} = 6x + 6$$

$\therefore \quad$ at the point where $x = 1$, $\quad \dfrac{d^2y}{dx^2} = 6(1) + 6 = +12$

This is positive and hence the turning point at $x = 1$ is a minimum.

The minimum value of y may be found by substituting $x = 1$ into the given equation. Hence

$$y_{min} = (1)^3 + 3(1)^2 - 9(1) + 6 = 1$$

At the point where $x = -3$, $\dfrac{d^2y}{dx^2} = 6(-3) + 6 = -12.$

This is negative and hence at $x = -3$ there is a maximum turning point. The maximum value of y may be found by substituting $x = -3$ into the given equation. Hence

$$y_{max} = (-3)^3 + 3(-3)^2 - 9(-3) + 6 = +3$$

To illustrate the test for maximum and minimum using the tangent gradient method, this method will be used to verify the above results.

At the turning point where $x = 1$, we know that

$$\frac{dy}{dx} = 0,$$

i.e. there is zero slope

and using a value of x slightly less than 1, say $x = 0.5$, gives

$$\frac{dy}{dx} = 3(0.5)^2 + 6(0.5) - 9 = -5.25$$

i.e. there is a negative slope

and using a value of x slightly greater than 1, say $x = 1.5$, gives

$$\frac{dy}{dx} = 3(1.5)^2 + 6(1.5) - 9 = +6.75$$

i.e. there is a positive slope

These results are best shown by means of a diagram (Fig. 3.5) which indicates clearly that when $x = 1$ we have a minimum.

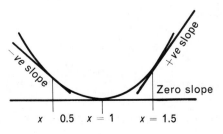

Fig. 3.5

Now at the turning point where $x = -3$ we know that

$$\frac{dy}{dx} = 0,$$

i.e. there is zero slope

and using a value of x slightly less than -3, say $x = -3.5$, gives

$$\frac{dy}{dx} = 3(-3.5)^2 + 6(-3.5) - 9 = +6.75$$

i.e. there is a positive slope

and using a value of x slightly greater than -3, say $x = -2.5$, gives

$$\frac{dy}{dx} = 3(-2.5)^2 + 6(-2.5) - 9 = -5.25$$

i.e. there is a negative slope

Fig. 3.6 indicates that when $x = -3$ we have a maximum turning point.

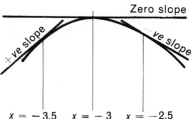

Fig. 3.6

APPLICATIONS IN TECHNOLOGY

There are many applications in technology which involve the finding of maxima and minima. The first step is to construct an equation connecting the quantity for which a maximum or minimum is required in terms of another variable. A diagram representing the problem may help in the formation of this initial equation.

EXAMPLE 3.3

A rectangular sheet of metal 360 mm by 240 mm has four equal squares cut out at the corners. The sides are then turned up to form a rectangular box. Find the length of the sides of the squares cut out so that the volume of the box may be as great as possible, and find this maximum volume.

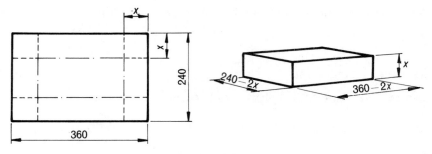

Fig. 3.7

Let the length of the side of each cut away square be x mm as shown in Fig. 3.7.

Hence the volume is

$$V = x(240 - 2x)(360 - 2x)$$
$$= 4x^3 - 1200x^2 + 86\,400x$$

$\therefore$
$$\frac{\mathrm{d}V}{\mathrm{d}x} = 12x^2 - 2400x + 86\,400$$

and
$$\frac{\mathrm{d}^2V}{\mathrm{d}x^2} = 24x - 2400$$

At a turning point
$$\frac{\mathrm{d}V}{\mathrm{d}x} = 0$$

$\therefore \qquad 12x^2 - 2400x + 86\,400 = 0$

or $\qquad x^2 - 200x + 7200 = 0 \quad$ by dividing through by 12

Now this is a quadratic equation which does not factorise so we will have to solve using the formula for the standard quadratic $ax^2 + bx + c = 0$ which gives $x = \dfrac{-b \pm \sqrt{b^2 - 4ac}}{2a}$.

Hence the solution of our equation is

$$x = \frac{-(-200) \pm \sqrt{(-200)^2 - 4 \times 1 \times 7200}}{2 \times 1}$$

$\therefore$ either $\qquad x = 152.9 \quad$ or $\quad x = 47.1$

However, from the physical sizes of the sheet, it is not possible for x to be 152.9 mm (since one side is only 240 mm long) so we reject this solution. Hence $x = 47.1$ mm.

Test for maximum or minimum:

From the above we have

$$\frac{d^2V}{dx^2} = 24x - 2400$$

and hence when $x = 47.1$

$$\frac{d^2V}{dx^2} = 24(47.1) - 2400 = -1270$$

This is negative, and hence V is a maximum when $x = 47.1$ mm.

It only remains to find the maximum volume by substituting $x = 47.1$ into the equation for V. Therefore

$$V_{max} = 47.1(240 - 2 \times 47.1)(360 - 2 \times 47.1)$$
$$= 1.825 \times 10^6 \text{ mm}^3$$

EXAMPLE 3.4

A cylinder with an open top has a capacity of $2\,\text{m}^3$ and is made from sheet metal. Neglecting any overlaps at the joints find the dimensions of the cylinder so that the amount of sheet steel used is a minimum.

Let the height of the cylinder be h metres and the radius of the base be r metres as shown in Fig. 3.8.

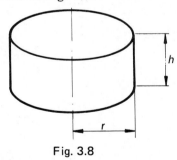

Fig. 3.8

Now the total area of metal = area of base + area of curved side

$$A = \pi r^2 + 2\pi r h$$

We cannot proceed to differentiate as there are two variables on the right-hand side of the equation. It is possible, however, to find a connection between r and h using the fact that the volume is $2\,\text{m}^3$.

Now   Volume of a cylinder $= \pi r^2 h$

$\therefore$ $2 = \pi r^2 h$

from which $h = \dfrac{2}{\pi r^2}$

We may now substitute for h in the equation for A.

$\therefore$ $A = \pi r^2 + 2\pi r \left(\dfrac{2}{\pi r^2}\right)$

$= \pi r^2 + \dfrac{4}{r}$

$= \pi r^2 + 4r^{-1}$

$\therefore$ $\dfrac{\mathrm{d}A}{\mathrm{d}r} = 2\pi r - 4r^{-2}$

and $\dfrac{\mathrm{d}^2 A}{\mathrm{d}r^2} = 2\pi + 8r^{-3}$

Now for a turning point

$\dfrac{\mathrm{d}A}{\mathrm{d}r} = 0$

or $2\pi r - 4r^{-2} = 0$

$\therefore$ $2\pi r - \dfrac{4}{r^2} = 0$

$\therefore$ $2\pi r = \dfrac{4}{r^2}$

$\therefore$ $r^3 = \dfrac{2}{\pi} = 0.637$

$r = \sqrt[3]{0.637} = 0.860$

To test for a minimum:

From above we have $\dfrac{\mathrm{d}^2 A}{\mathrm{d}r^2} = 2\pi + 8r^{-3}$

$= 2\pi + \dfrac{8}{r^3}$

We do not need to do any further calculation here as this expression must be positive for all positive values of r. Hence $r = 0.86$ makes A a minimum.

We may find the corresponding value of h by substituting $r = 0.86$ into the equation found previously for h in terms of r.

$$h = \frac{2}{\pi(0.86)^2} = 0.86$$

Hence for the minimum amount of metal to be used the radius is 0.86 m and the height is 0.86 m.

Exercise 3.1

1) Find the maximum and minimum values of:

(a) $y = 2x^3 - 3x^2 - 12x + 4$;

(b) $y = x^3 - 3x^2 + 4$;

(c) $y = 6x^2 + x^3$

2) Given that $y = 60x + 3x^2 - 4x^3$, calculate:

(a) the gradient of the tangent to the curve of y at the point where $x = 1$;

(b) the value of x for which y has its maximum value;

(c) the value of x for which y has its minimum value.

3) Calculate the coordinates of the points on the curve

$$y = x^3 - 3x^2 - 9x + 12$$

at each of which the tangent to the curve is parallel to the x-axis.

4) A curve has the equation $y = 8 + 2x - x^2$. Find:

(a) the value of x for which the gradient of the curve is 6;

(b) the value of x which gives the maximum value of y;

(c) the maximum value of y.

5) The curve $y = 2x^2 + \dfrac{k}{x}$ has a gradient of 5 when $x = 2$.

Calculate:

(a) the value of k;

(b) the minimum value of y.

6) From a rectangular sheet of metal measuring 120 mm by 75 mm equal squares of side x are cut from each of the corners. The remaining flaps are then folded upwards to form an open box. Prove that the volume of the box is given by

$$V = 9000x - 390x^2 + 4x^3$$

Find the value of x such that the volume is a maximum.

7) An open rectangular tank of height h metres with a square base of side x metres is to be constructed so that it has a capacity of 500 cubic metres. Prove that the surface area of the four walls and the base will be $\dfrac{2000}{x} + x^2$ square metres. Find the value of x for this expression to be a minimum.

8) The volume of a cone is given by the formula $V = \frac{1}{3}\pi r^2 h$, where h is the height of the cone and r its radius. If $h = 6 - r$, calculate the value of r for which the volume is a maximum.

9) A box without a lid has a square base of side x mm and rectangular sides of height h mm. It is made from $10\,800$ mm² of sheet metal of negligible thickness. Prove that $h = \dfrac{10\,800 - x^2}{4x}$ and that the volume of the box is $(2700x - \frac{1}{4}x^3)$. Hence calculate the maximum volume of the box.

10) A cylindrical tank, with an open top, is to be made to hold 300 cubic metres of liquid. Find the dimensions of the tank so that its surface area shall be a minimum.

11) A cooling tank is to be made with the trapezoidal section as shown:

Its cross-sectional area is to be $300\,000$ mm². Show that the width of material needed to form, from one sheet, the bottom and folded-up sides is $w = \dfrac{300\,000}{h} + 1.828h$. Hence find the height h of the tank so that the width of material needed is a minimum.

12) A cylindrical cup is to be drawn from a disc of metal of 50 mm diameter. Assuming that the surface area of the cup is the same as that of the disc find the dimensions of the cup so that its volume is a maximum.

13) A lever weighing $12\,\text{N}$ per m run of its length is as shown:

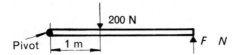

Find the length of the lever so that the force F shall be a minimum.

14) The cost per hour of running a certain machine is

$$C = 1.20 + 0.06\,N^3$$

where N is the number of components produced per hour. Find the most economical value of N if 1000 components are to be produced.

15) A rectangle is inscribed in a circle of $120\,\text{mm}$ diameter. Show that the rectangle having the largest area is a square, and find the length of its side.

16) The efficiency of a steam turbine is given by

$$\eta = 4(n\rho \cos\alpha - n^2\rho^2)$$

where n and α are constants. Find the maximum value of η.

4. INTEGRATION

On reaching the end of this chapter you should be able to:

1. *Determine indefinite integrals of functions involving sin ax, cos ax and e^{ax}.*
2. *Evaluate the definite integrals involving sin ax, cos ax and e^{ax}.*
3. *Define the mean and root mean square values of functions over a given range.*
4. *Evaluate the mean and root mean square values of simple periodic functions.*

INTRODUCTION

The table of differential coefficients on p. 15 shows that by differentiating $\sin ax$ with respect to x we obtain $a \cos ax$. Hence by differentiating $\dfrac{1}{a} \sin ax$ with respect to x we obtain $\cos ax$.

Now since integration is the reverse of differentiation, it follows that by integrating $\cos ax$ with respect to x we obtain $\dfrac{1}{a} \sin ax$.

Now by making small modifications similar to the one just described, we may rewrite the table on p. 15 as a table showing integrals of the more common functions:

y	$\int y \, dx$
ax^n	$\dfrac{a}{n+1} x^{n+1}$
$\sin ax$	$-\dfrac{1}{a} \cos ax$
$\cos ax$	$\dfrac{1}{a} \sin ax$
$\sec^2 x$	$\tan x$
$\dfrac{1}{x}$	$\log_e x$
e^{ax}	$\dfrac{1}{a} e^{ax}$

INDEFINITE INTEGRALS

An indefinite integral is an integral without *limits* and the solution must therefore contain a constant of integration.

EXAMPLE 4.1

Find $\int (x^2 + 2x - 3)\, dx$.

$$\int (x^2 + 2x - 3)\, dx \;=\; \frac{x^3}{3} + 2\frac{x^2}{2} - 3x + k$$

$$=\; \frac{x^3}{3} + x^2 - 3x + k$$

EXAMPLE 4.2

Find $\int (\sin 7\theta + 2 \cos 5\theta)\, d\theta$.

$$\int (\sin 7\theta + 2 \cos 5\theta)\, d\theta \;=\; -\tfrac{1}{7}\cos 7\theta + \tfrac{2}{5}\sin 5\theta + k$$

EXAMPLE 4.3

Find $\int \left(e^{6t} - \dfrac{1}{e^{3t}} \right) dt$.

$$\int \left(e^{6t} - \frac{1}{e^{3t}} \right) dt \;=\; \int (e^{6t} - e^{-3t})\, dt$$

$$=\; \tfrac{1}{6} e^{6t} - \frac{1}{(-3)} e^{-3t} + k$$

$$=\; \tfrac{1}{6} e^{6t} + \tfrac{1}{3} e^{-3t} + k$$

DEFINITE INTEGRALS

A definite integral has limits. Square brackets must be used which indicate that 'we have completed the actual integration and will next be substituting the values of the limits'.

EXAMPLE 4.4

Evaluate $\int_2^3 (1 + \cos 2\phi)\, d\phi$.

We should remember that when limits are substituted into trigono-metrical functions they represent radian values (not degrees).

$$\int_2^3 (1 + \cos 2\phi)\, d\phi = [\phi + \tfrac{1}{2} \sin 2\phi]_2^3$$

$$= \{3 + \tfrac{1}{2} \sin (2 \times 3)\} - \{2 + \tfrac{1}{2} \sin (2 \times 2)\}$$

$$= 3 + \tfrac{1}{2} \sin 6 - 2 - \tfrac{1}{2} \sin 4$$

$$= 1.24$$

EXAMPLE 4.5

Evaluate $\int_0^1 5(e^{2t} - e^{-2t})\, dt$.

$$\int_0^1 5(e^{2t} - e^{-2t})\, dt = 5 \int_0^1 (e^{2t} - e^{-2t})\, dt$$

$$= 5 \left[\tfrac{1}{2} e^{2t} - \frac{1}{(-2)} e^{-2t} \right]_0^1$$

$$= \tfrac{5}{2} \{(e^{2 \times 1} + e^{-2 \times 1}) - (e^{2 \times 0} + e^{-2 \times 0})\}$$

$$= \tfrac{5}{2} \{e^2 + e^{-2} - e^0 - e^0\}$$

$$= \tfrac{5}{2} \{7.39 + 0.14 - 1 - 1\} = 13.8$$

EXAMPLE 4.6

Evaluate $\int_1^2 \left(\frac{1}{x} + \sec^2 x \right) dx$.

$$\int_1^2 \left(\frac{1}{x} + \sec^2 x \right) dx = [\log_e x + \tan x]_1^2$$

$$= \{\log_e 2 + \tan 2\} - \{\log_e 1 + \tan 1\}$$

$$= 0.693 + (-2.185) - 0 - 1.557$$

$$= -3.05$$

Exercise 4.1

Find:

1) $\int \left(5x^2 + 2x - \dfrac{4}{x^2}\right) dx$

2) $\int \left(\sqrt{x} + \dfrac{1}{\sqrt{x}}\right) dx$

3) $\int \sin \dfrac{x}{3} \, dx$

4) $\int 5 \cos 3\theta \, d\theta$

5) $\int (1 + \sin \frac{2}{3}\phi) \, d\phi$

6) $\int \left(\cos \dfrac{\theta}{2} - \sin \dfrac{3\theta}{2}\right) d\theta$

7) $\int (2t + \sin 2t) \, dt$

8) $\int e^{3x} \, dx$

9) $\int e^{-0.5u} \, du$

10) $\int (3e^{2t} - 2e^t) \, dt$

11) $\int (e^{-x/2} + e^{3x/2}) \, dx$

12) $\int \left(\sec^2 x + \dfrac{1}{x}\right) dx$

Evaluate:

13) $\displaystyle\int_1^2 (x^3 + 4) \, dx$

14) $\displaystyle\int_0^2 \left(\dfrac{x^2 + x^3}{x}\right) dx$

15) $\displaystyle\int_0^1 \dfrac{2 \cos x}{3} \, dx$

16) $\displaystyle\int_0^{\pi/2} 3 \sin 4\phi \, d\phi$

17) $\displaystyle\int_{\pi/6}^{\pi/3} 2 \sin \dfrac{2t}{3} \, dt$

18) $\displaystyle\int_{-\pi/2}^{\pi/2} \sin 2\theta \, d\theta$

19) $\displaystyle\int_{\pi/2}^{\pi} \cos \dfrac{\phi}{2} \, d\phi$

20) $\displaystyle\int_{-0.2}^{0.5} (1 + 0.6 \cos 0.2\theta) \, d\theta$

21) $\displaystyle\int_0^{\pi} (\sin x - \sin 3x) \, dx$

22) $\displaystyle\int_0^1 e^x \, dx$

23) $\displaystyle\int_1^2 e^{-2t} \, dt$

24) $\displaystyle\int_{0.5}^1 \left(e^{\theta/3} - \dfrac{1}{e^{\theta/3}}\right) d\theta$

25) $\displaystyle\int_1^2 (1 + 2e^{0.3v}) \, dv$

26) $\displaystyle\int_2^3 4(e^u + e^{-u}) \, du$

27) $\displaystyle\int_0^3 2e^{-0.4t}\, dt$ **28)** $\displaystyle\int_{-1}^1 \frac{4}{5e^{1.4x}}\, dx$

29) $\displaystyle\int_2^3 \frac{2}{x}\, dx$ **30)** $\displaystyle\int_{\pi/4}^{\pi/3} \sec^2\theta\, d\theta$

MEAN VALUE

The mean (or average) value or height of a curve is often of importance.

$$\text{The mean value} = \frac{\text{Area under the curve}}{\text{Length of the base}}$$

A type of graph which is met frequently in technology is a waveform.

A waveform is a graph which repeats indefinitely. A sine curve (Fig. 4.1) is an example of a waveform.

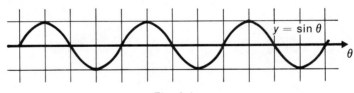

Fig. 4.1

A portion of the graph which shows the complete shape of the waveform without any repetition is called a cycle.

In the case of the curve of sin θ the portion of the graph over one cycle is said to be a *full wave* (Fig. 4.2), and over half of one cycle is said to be a *half wave* (Fig. 4.3).

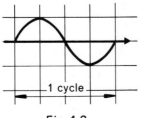

Fig. 4.2

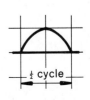

Fig. 4.3

EXAMPLE 4.7

Find the mean value of $A \sin \theta$ for:

a) a half wave   **b)** a full wave

a) A half wave of $A \sin \theta$ occurs over a range from $\theta = 0$ rad to $\theta = \pi$ rad as shown in Fig. 4.4.

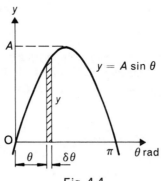

Fig. 4.4

$$\text{Area of elementary strip} = y \, \delta\theta$$

$\therefore$ $$\text{Total area under curve} = \sum_{\theta = 0}^{\theta = \pi} y \, \delta\theta \quad \text{approximately}$$

$$= \int_0^\pi y \, d\theta \quad \text{exactly}$$

$$= \int_0^\pi A \sin \theta \, d\theta$$

$$= A \int_0^\pi \sin \theta \, d\theta$$

$$= A[-\cos]_0^\pi$$

$$= A\{(-\cos \pi) - (-\cos 0)\}$$

$$= A\{-(-1) - (-1)\}$$

$$= 2A \text{ square units}$$

Now $$\text{Mean value} = \frac{\text{Area under curve}}{\text{Length of base}}$$

$$= \frac{2A}{\pi} = 0.637A$$

b) A full wave of $A \sin \theta$ occurs over a range from $\theta = 0$ rad to $\theta = 2\pi$ rad as shown in Fig. 4.5.

The method used is similar to that in part **a)** except that the limits of the integral are now 0 and 2π.

$\therefore$

$$\text{Total area under curve} = \int_0^{2\pi} A \sin \theta \, d\theta$$

$$= A[-\cos \theta]_0^{2\pi}$$

$$= 0$$

This zero answer is because the area under the second half of the wave is calculated as negative which is added to a similar positive area under the first half wave. Thus the mean value is also zero.

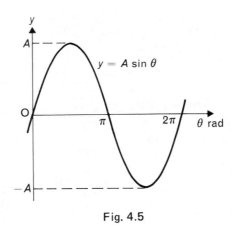

Fig. 4.5

EXAMPLE 4.8

A ramp waveform consists of a series of right-angled triangles as shown in Fig. 4.6. Find the mean height of the waveform.

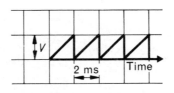

Fig. 4.6

The triangular area under the graph is $\frac{1}{2} \times 2 \times V = V$. However it is instructive to find the area by integration as a similar method is used later for finding root mean squares.

We must first find an equation for the sloping side of the triangle which is shown set up on v–t axes as shown in Fig. 4.7.

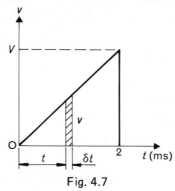

Fig. 4.7

The equation is

$$v = (\text{gradient}) t$$

or

$$v = \left(\frac{V}{2}\right) t$$

Now

The triangular area = Sum of elementary strip areas

$$= \sum_{t=0}^{t=2} v \, \delta t \quad \text{approximately}$$

$$= \int_0^2 v \, dt \quad \text{exactly}$$

$$= \int_0^2 \left(\frac{V}{2}\right) t \, dt$$

$$= \frac{V}{2} \int_0^2 t \, dt = \frac{V}{2} \left[\frac{t^2}{2}\right]_0^2 = V$$

Now   Mean value $= \dfrac{\text{Area under graph}}{\text{Length of base}}$

$$= \frac{V}{2}$$

ROOT MEAN SQUARE (r.m.s.) VALUE

In alternating current work the mean value is not of great importance. This is because we are usually interested in the power produced and this depends on the square of the current or voltage values. In these cases we use the root mean square value.

The root mean square value $= \sqrt{\text{Average height of the } y^2 \text{ curve}}$

i.e.   r.m.s. $= \sqrt{\dfrac{\text{Area under the } y^2 \text{ curve}}{\text{Length of the base}}}$

EXAMPLE 4.9

Find the r.m.s. value for $A \sin \theta$ for:

a) a half wave   b) a full wave

a) The method is similar to that used in Example 4.7 except that we use y^2 instead of y in the integral.

Thus

$$\text{Total area under } y^2 \text{ curve} = \int_0^\pi y^2 \, d\theta$$

$$= \int_0^\pi (A \sin \theta)^2 \, d\theta$$

$$= A^2 \int_0^\pi \sin^2\theta \, d\theta$$

However $\sin^2\theta = \tfrac{1}{2}(1 - \cos 2\theta)$ as on p. 205.

Therefore

$$\text{Total area under } y^2 \text{ curve} = \frac{A^2}{2} \int_0^\pi (1 - \cos 2\theta) \, d\theta$$

$$= \frac{A^2}{2} [\theta - \tfrac{1}{2} \sin 2\theta]_0^\pi$$

$$= \frac{A^2}{2} \{(\pi - \tfrac{1}{2} \sin 2 \times \pi) - (0 - \tfrac{1}{2} \sin 2 \times 0)\}$$

$$= \frac{A^2}{2} \{\pi - \tfrac{1}{2} \times 0 - 0 + \tfrac{1}{2} \times 0\}$$

$$= \frac{\pi A^2}{2}$$

Now r.m.s. value $= \sqrt{\dfrac{\text{Area under } y^2 \text{ curve}}{\text{Length of base}}}$

$$= \sqrt{\dfrac{\pi A^2/2}{\pi}}$$

$$= \dfrac{A}{\sqrt{2}} = 0.707\,A$$

b) For a full wave the working is similar to that in part **a)** except that the limits are 0 to 2π rad. The reader may find it useful to verify that the same result of $0.707A$ is obtained.

Similar results are also obtained for cosine waveforms. Thus:

> for sinusoidal waveforms the r.m.s. value is 0.707 of the amplitude or peak value.

EXAMPLE 4.10

Find the r.m.s. value of the ramp waveform shown in Fig. 4.7.

The method is similar to that used in Example 4.8 except that we use v^2 instead of v in the integral.

Thus

$$\text{Total area under the } v^2 \text{ graph} = \int_0^2 v^2 \, dt$$

$$= \int_0^2 \left\{ \left(\dfrac{V}{2}\right)t \right\}^2 dt$$

$$= \dfrac{V^2}{4} \int_0^2 t^2 \, dt$$

$$= \dfrac{V^2}{4} \left[\dfrac{t^3}{3} \right]_0^2$$

$$= \dfrac{V^2}{4} \left(\dfrac{2^3}{3} - \dfrac{0^3}{3} \right)$$

$$= \dfrac{2V^2}{3}$$

Now $\qquad$ r.m.s. value $= \sqrt{\dfrac{\text{Area under } v^2 \text{ graph}}{\text{Length of base}}}$

$$= \sqrt{\dfrac{2V^2/3}{2}}$$

$$= \dfrac{V}{\sqrt{3}} = 0.577V$$

Exercise 4.2

1) Find the mean value of $V \cos \theta$ for:

(a) a half cycle;

(b) a cycle.

2) Find the mean value of $6 \sin 2\theta$ for:

(a) a half cycle;

(b) a cycle.

Find the mean values of the waveforms shown in Questions 3), 4), 5) and 6) over one cycle in each case.

3)

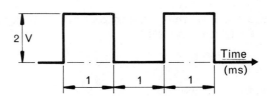

4)

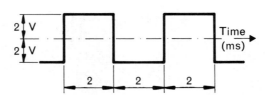

5)

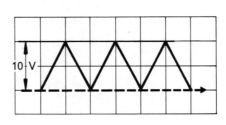

6)

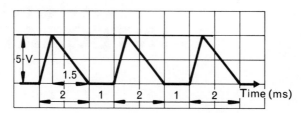

7) Find the mean value of the waveform whose shape for one cycle is as shown.

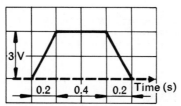

8) Find the r.m.s. value of $V \cos \theta$ over:

(a) a half wave;

(b) a full wave.

9) Find the r.m.s. value of $6 \sin 2\theta$ over:

(a) a half cycle;

(b) a cycle.

10) Find the r.m.s. values of the waveforms in Questions 3), 4), 5), 6) and 7) over one cycle in each case.

5. NUMERICAL INTEGRATION

On reaching the end of this chapter you should be able to:

1. Derive the trapezoidal and mid-ordinate rules for numerical integration.
2. Derive Simpson's rule over two intervals.
3. Deduce the general form of Simpson's rule over an even number of equal intervals.
4. Evaluate definite integrals using the trapezoidal, mid-ordinate and Simpson's rules.
5. Determine answers to a desired accuracy using Simpson's rule.

INTRODUCTION

We know that the area under a curve between limits $x = a$ and $x = b$ is given by the sum of all the elementary strip areas (Fig. 5.1).

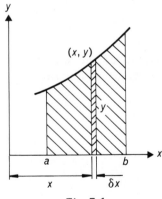

Fig. 5.1

In mathematical notation

$$\text{Area} = \sum_{x=a}^{x=b} y\, \delta x \quad \text{approximately}$$

or

$$\text{Area} = \int_a^b y\, dx \quad \text{exactly}$$

However, it is not always possible to evaluate the integral by direct mathematical integration. For example, we may have a curve obtained from experimental results for which there is no equation giving y in terms of x. Another difficulty arises when the expression for y in terms of x is too complicated for us to integrate.

We may then have to resort to dividing the required area into a number of strips (similar to elementary strips), finding the area of each strip, and then adding these up. Therefore the result will depend on calculating, as accurately as possible, the areas of the vertical strips.

Three reasonably simple methods are available: trapezoidal, mid-ordinate and Simpson's rules. You may have used these previously, but we shall now examine, in more detail, their derivation and the accuracy of results obtained.

The worked examples, which appear later in this chapter, to illustrate the use of the three methods can also be solved by direct integration and thus exact results obtained. You may find it useful to check these. This enables the error to be calculated for each result.

TRAPEZOIDAL RULE

Consider the area having the boundary ABCD shown in Fig. 5.2.

The area is divided into a number of vertical strips of equal width b.

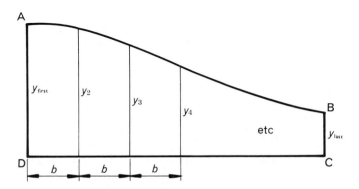

Fig. 5.2

Each vertical strip is assumed to be a trapezium. Hence the third strip, for example, will have an area $= b \times \frac{1}{2}(y_3 + y_4)$.

But

Area ABCD $=$ The sum of all the vertical strips

$$= b \times \tfrac{1}{2}(y_{\text{first}} + y_2) + b \times \tfrac{1}{2}(y_2 + y_3) + b \times \tfrac{1}{2}(y_3 + y_4) + \ldots$$

$$= b[\tfrac{1}{2}y_{\text{first}} + \tfrac{1}{2}y_2 + \tfrac{1}{2}y_2 + \tfrac{1}{2}y_3 + \ldots + \tfrac{1}{2}y_{\text{last}}]$$

$$= b[\tfrac{1}{2}y_{\text{first}} + y_{\text{last}}) + y_2 + y_3 + y_4 \ldots]$$

Area ABCD $=$ Strip width $\times [\tfrac{1}{2}$(Sum of first and last ordinates)
$\qquad\qquad\qquad\qquad$ + (Sum of remaining ordinates)]

EXAMPLE 5.1

Find $\displaystyle\int_0^{1.5} e^x \, dx$ by numerical integration using the trapezoidal rule.

The integral represents the area under the curve $y = e^x$ as shown in Fig. 5.3. The area has been divided into six vertical strips of equal width and the lengths of the ordinates found using a calculator.

Now the trapezoidal rule states

$$\text{Area} = (\text{Strip width}) \times [\tfrac{1}{2}(\text{Sum of first and last ordinates})$$
$$+ (\text{Sum of remaining ordinates})]$$

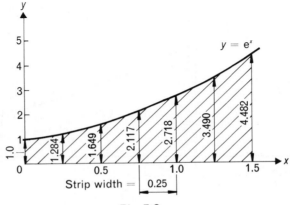

Fig. 5.3

Thus

$$\int_0^{1.5} e^x \, dx = \text{Area under the curve}$$

$$\simeq 0.25 \times [\tfrac{1}{2}(1.000 + 4.482) + (1.284 + 1.649 + 2.117$$
$$+ 2.718 + 3.490)]$$

$$= 3.500 \quad \text{correct to 3 decimal places}$$

The result by integration is 3.482 correct to 3 decimal places.

$$\therefore \quad \text{Error} = \frac{3.500 - 3.482}{3.482} \times 100 = +0.52\%$$

EXAMPLE 5.2

Find $\int_0^\pi \sin \theta \, d\theta$ by numerical integration using the trapezoidal rule.

The integral represents the area under the curve $y = \sin \theta$ as shown in Fig. 5.4. The area has been divided into six vertical strips of equal width and the lengths of the ordinates found using a calculator.

Now the trapezoidal rule states

$$\text{Area} = (\text{Strip width}) \times [\tfrac{1}{2}(\text{Sum of first and last ordinates})$$
$$+ (\text{Sum of remaining ordinates})]$$

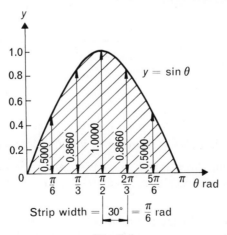

Fig. 5.4

Thus

$$\int_0^\pi \sin\theta \; d\theta = \text{Area under the curve}$$

$$\simeq \frac{\pi}{6} \times [\tfrac{1}{2}(0+0) + (0.5000 + 0.8660 + 1.0000$$

$$+ 0.8660 + 0.5000)]$$

$$= 1.954 \quad \text{correct to 3 decimal places}$$

The result by integration is 2.000 correct to 3 decimal places.

$$\therefore \qquad \text{Error} = \frac{1.954 - 2.000}{2.000} \times 100 = -2.3\%$$

MID-ORDINATE RULE

Consider the area having the boundary ABCD shown in Fig. 5.5.

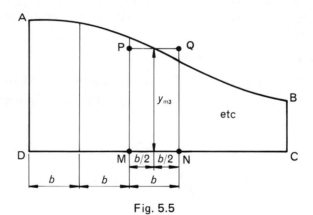

Fig. 5.5

The area is divided into a number of vertical strips of equal width b.

In the third strip the vertical line of length y_{m3} is half way between the boundary ordinates and is called the mid-ordinate of the third strip.

The area of the third strip is assumed to be the same as the area of the rectangle PQNM which is MN $\times$ PM or by_{m3}.

Thus, if the mid-ordinates of all the strips are y_{m1}, y_{m2}, y_{m3}, etc.

then   Area ABCD $= by_{m1} + by_{m2} + by_{m3} + \ldots$

$$= b(y_{m1} + y_{m2} + y_{m3} + \ldots y_{m\,last})$$

or $\boxed{\text{Area ABCD} = (\text{Strip width}) \times (\text{Sum of mid-ordinates})}$

EXAMPLE 5.3

Find $\displaystyle\int_0^{1.5} e^x\,dx$ by numerical integration using the mid-ordinate rule.

The integral represents the area under the curve $y = e^x$ as shown in Fig. 5.6. The area has been divided into six vertical strips of equal width and the lengths of the mid-ordinates found using a calculator.

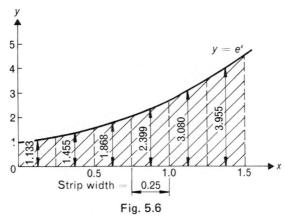

Fig. 5.6

Now the mid-ordinate rule states

$$\text{Area} = (\text{Strip width}) \times (\text{Sum of mid-ordinates})$$

Thus

$$\int_0^{1.5} e^x\,dx = \text{Area under the curve}$$

$$\simeq 0.25 \times (1.133 + 1.455 + 1.868 + 2.399 + 3.080 + 3.955)$$

$$= 3.473 \quad \text{correct to 3 decimal places}$$

The result by integration is 3.482 correct to 3 decimal places.

$$\therefore \qquad \text{Error} = \frac{3.473 - 3.482}{3.482} \times 100 = -0.26\%$$

EXAMPLE 5.4

Find $\int_0^\pi \sin\theta \, d\theta$ by numerical integration using the mid-ordinate rule.

The integral represents the area under the curve $y = \sin\theta$ as shown in Fig. 5.7. The area has been divided into six vertical strips of equal width and the lengths of the mid-ordinates found using a calculator.

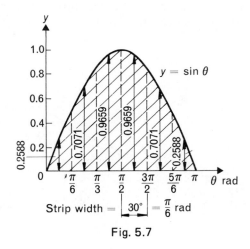

Fig. 5.7

Now the mid-ordinate rule states

$$\text{Area} = (\text{Strip width}) \times (\text{Sum of mid-ordinates})$$

Thus

$$\int_0^\pi \sin\theta \, d\theta = \text{Area under the curve}$$

$$\simeq \frac{\pi}{6} \times (0.2588 + 0.7071 + 0.9659 + 0.9659$$

$$+ \, 0.7071 + 0.2588)$$

$$= 2.023 \quad \text{correct to 3 decimal places}$$

The result by integration is 2.000 correct to 3 decimal places.

$$\therefore \quad \text{Error} = \frac{2.023 - 2.000}{2.000} \times 100 = +1.15\%$$

SIMPSON'S RULE

Suppose we wish to find the area PQRNL shown shaded in Fig. 5.8. QM is the mid-ordinate which divides the area into two strips of equal width, b. Fig. 5.9 is a repeat of Fig. 5.8 but shows a straight line AB drawn through point Q and parallel to chord PR.

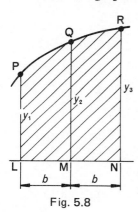

Fig. 5.8

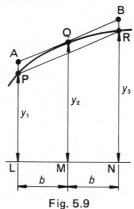

Fig. 5.9

Now

$$\begin{pmatrix} \text{Area of} \\ \text{parallelogram ABRP} \end{pmatrix} = \begin{pmatrix} \text{Area of} \\ \text{trapezoid ABNL} \end{pmatrix} - \begin{pmatrix} \text{Area of} \\ \text{trapezoid PRNL} \end{pmatrix}$$

$$= 2by_2 - \tfrac{1}{2}(y_1 + y_3)2b$$

$$= b(2y_2 - y_1 - y_3)$$

Now the area PQR, between the curve and the chord, is taken as $\tfrac{2}{3}$ of the area of parallelogram ABRP. This is a very close approximation. It is only exact if the curve is the arc of a parabola and also AB is the tangent at Q.

$$\text{Thus} \quad \text{Area PQRNL} = \begin{pmatrix} \text{Area of} \\ \text{trapezoid PRNL} \end{pmatrix} + \text{Area PQR}$$

$$= \tfrac{1}{2}(y_1 + y_3)2b + \tfrac{2}{3}b(2y_2 - y_1 - y_3)$$

$$= b(y_1 + y_3 + \tfrac{4}{3}y_2 - \tfrac{2}{3}y_1 - \tfrac{2}{3}y_3)$$

$$= \frac{b}{3}(y_1 + 4y_2 + y_3)$$

Fig. 5.10 shows a larger area divided into six vertical strip areas of equal width. Each adjacent pair of strips forms an area similar to that shown in Fig. 5.8. We therefore apply the expression we derived for each pair of strips as follows:

$$\text{Shaded area} = \begin{pmatrix} \text{Area between} \\ \text{ordinates} \\ y_1 \text{ and } y_3 \end{pmatrix} + \begin{pmatrix} \text{Area between} \\ \text{ordinates} \\ y_3 \text{ and } y_5 \end{pmatrix} + \begin{pmatrix} \text{Area between} \\ \text{ordinates} \\ y_5 \text{ and } y_7 \end{pmatrix}$$

$$= \frac{b}{3}(y_1 + 4y_2 + y_3) + \frac{b}{3}(y_3 + 4y_4 + y_5)$$

$$+ \frac{b}{3}(y_5 + 4y_6 + y_7)$$

$$= \tfrac{1}{3}b[y_1 + y_7 + 4(y_2 + y_4 + y_6) + 2(y_3 + y_5)]$$

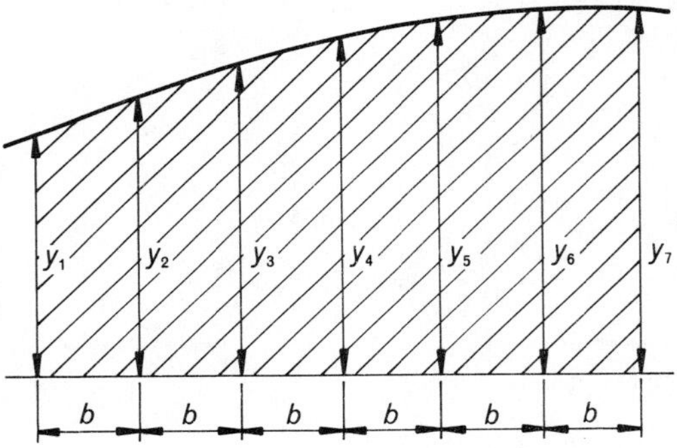

Fig. 5.10

This idea may be extended providing there is an *even* number of strips of equal width to give a general expression for Simpson's rule:

$$\text{Area} = \tfrac{1}{3}(\text{Strip width}) \times [(\text{Sum of first and last ordinates})$$
$$+ 4(\text{Sum of even ordinates})$$
$$+ 2(\text{Sum of remaining odd ordinates})]$$

EXAMPLE 5.5

Find $\int_0^{1.5} e^x \, dx$ by numerical integration using Simpson's rule.

In order to use Simpson's rule the required area must be divided into an *even* number of strips. This has been done in Fig. 5.3 and we will use the values of the ordinates given.

Now Simpson's rule states

Area $= \frac{1}{3}$(Strip width) $\times$ [(Sum of first and last ordinates)

$\qquad\qquad\qquad + 4$(Sum of even ordinates)

$\qquad\qquad\qquad + 2$(Sum of remaining odd ordinates)]

Thus

$\int_0^{1.5} e^x \, dx =$ Area under the curve

$\qquad \simeq \frac{1}{3} \times 0.25 \times [(1.000 + 4.482)$

$\qquad\qquad\qquad + 4(1.284 + 2.117 + 3.490$

$\qquad\qquad\qquad + 2(1.649 + 2.718)]$

$\qquad = 3.482 \quad$ correct to 3 decimal places

The result by integration is 3.482 correct to 3 decimal places and so there is no error.

An alternative layout using a table to show the calculations is often used:

Ordinate number	Length	Simpson's multiplier	Product
1	1.000	1	1.000
2	1.284	4	5.136
3	1.649	2	3.298
4	2.117	4	8.468
5	2.718	2	5.436
6	3.490	4	13.960
7	4.482	1	4.482
		Total product = 41.780	

Hence $\qquad$ Result $= \frac{1}{3} \times 0.25 \times 41.780 = 3.482$

EXAMPLE 5.6

Find $\int_0^{\pi} \sin \theta \, d\theta$ by numerical integration using Simpson's rule.

In order to use Simpson's rule the required area must be divided into an *even* number of strips. This has been done in Fig. 5.4 and we will use the values of the ordinates given.

Now Simpson's rule states

Area $= \frac{1}{3}$(Strip width) $\times$ [(Sum of first and last ordinates)
$+ 4$(Sum of even ordinates)
$+ 2$(Sum of remaining odd ordinates)]

Thus

$\int_0^\pi \sin\theta \; d\theta =$ Area under the curve

$\simeq \dfrac{1}{3} \times \dfrac{\pi}{6} \times [(0+0) + 4(0.5000 + 1.0000 + 0.5000)$

$+ 2(0.8660 + 0.8660)]$

$= 2.001$ correct to 3 decimal places

The result by integration is 2.000 correct to 3 decimal places.

$\therefore$ Error $= \dfrac{2.001 - 2.000}{2.000} \times 100 = 0.05\%$

COMPARISON OF TRAPEZOIDAL, MID-ORDINATE AND SIMPSON'S RULES

If a curve is shaped ⌒ then the trapezoidal rule gives the shaded area shown in Fig. 5.11. This area is smaller than the true area under the curve, and this is confirmed by the answer to Example 5.2 which is 2.3% too small.

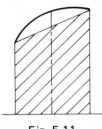

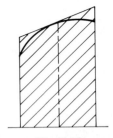

Fig. 5.11   Fig. 5.12

The mid-ordinate rule gives the shaded area shown in Fig. 5.12, which is greater than the true area under the curve. This is confirmed by the answer to Example 5.4 which is 1.15% too large.

For a curve shaped $\smile$ the trapezoid rule gives results too large (see Example 5.1), whilst the mid-ordinate rule gives results too small (see Example 5.3).

Simpson's rule, which is a combination of both the other rules, is much more accurate as confirmed by the results of Examples 5.5 and 5.6. Hence if accuracy is of importance then Simpson's rule should be used.

ACCURACY OF ANSWERS USING SIMPSON'S RULE

In the preceding text it was possible to obtain exact answers to the worked examples. This enabled us to compare results obtained using approximate methods so that we may gain some idea of the degree of accuracy which may be expected.

However, in practice, we would only use Simpson's rule for integration if it was *not* possible to get an exact answer.

The accuracy of a result using Simpson's rule will depend on the number of intervals chosen. There are expressions which may be used to estimate the error, and also to decide on the number of intervals required to obtain an answer to a particular degree of accuracy. However these are tedious to use and it is recommended that the following method be used:

> If an increase in the number of equal intervals does not involve a change in the answer to a certain degree of of accuracy, then we may rely on the result to that degree of accuracy.

Suppose, for example, that for a particular problem when using Simpson's rule we obtain (correct to 3 significant figures):

> with   6 equal intervals an answer of 6.36
>
> with   8 equal intervals an answer of 6.34
>
> with 10 equal intervals an answer of 6.33

and   with 12 equal intervals an answer of 6.33

Here we may safely assume the result as 6.33 correct to 3 significant figures.

Exercise 5.1

Evaluate, stating the answers to 3 significant figures, the following integrals:

(a) the trapezoidal rule;

(b) the mid-ordinate rule;

(c) Simpson's rule.

1) $\int_0^3 \sqrt{(9-x^2)}\, dx$ using 6 intervals

2) $\int_0^{1.5} \dfrac{dx}{\sqrt{(4-x^2)}}$ using 6 intervals

3) $\int_1^2 (\log_e x)\, dx$ using 10 intervals

4) $\int_0^{\pi/4} \left(\dfrac{\theta}{\cos\theta}\right) d\theta$ using 6 intervals

5) $\int_{0.1}^{0.7} \left(\dfrac{t}{1-t}\right) dt$ using 6 intervals

6) $\int_0^{\pi/6} (\sin^3\phi)\, d\phi$ using 6 intervals

7) $\int_0^2 (x^2\, e^{-x})\, dx$ using 8 intervals

8) Use Simpson's rule to find the value of

$$\int_{1.8}^3 \left(\dfrac{1}{\sin x}\right) dx$$

giving the answer *correct* to 3 significant figures. Start with 6 equal intervals and proceed until the required accuracy is guaranteed.

6.

DIFFERENTIAL EQUATIONS

FAMILIES OF CURVES

Suppose we know that $\qquad \dfrac{dy}{dx} = 3$

This may be rewritten as $\qquad y = \displaystyle\int 3\,dx$

from which $\qquad y = 3x + C$

The constant of integration C represents any number.

Now suppose we give C different values and plot the graph of each equation (Fig. 6.1).

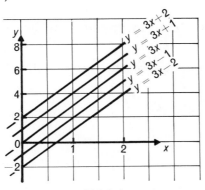

Fig. 6.1

71

When $\qquad C = -2: \qquad y = 3x - 2$

and when $\qquad C = -1: \qquad y = 3x - 1$

and when $\qquad C = 0: \qquad y = 3x$

and when $\qquad C = +1: \qquad y = 3x + 1$

and when $\qquad C = +2: \qquad y = 3x + 2$

It follows that the equation $y = 3x + C$ represents a set of graphs called a *family*. They all have one thing in common, that is $\dfrac{dy}{dx} = 3$.

If we specify a particular point we may find the equation of the graph which passes through that point.

EXAMPLE 6.1

Find the equation of the graph which is one of the family represented by the equation $y = 3x + C$ and which passes through the point $(1, 7)$.

If a point lies on a graph its coordinates satisfy the equation of the graph.

Hence since the point $(1, 7)$ lies on the graph $y = 3x + C$

then $\qquad\qquad\qquad 7 = 3(1) + C$

from which $\qquad\qquad C = 4$

and therefore the required equation is $y = 3x + 4$.

EXAMPLE 6.2

Sketch the family of curves represented by the equation $\dfrac{dy}{dx} = 2x + 1$.

Find also the equation of the curve which passes through the point $(3, 18)$.

We have $\qquad\qquad\qquad \dfrac{dy}{dx} = 2x + 1$

or $\qquad\qquad\qquad y = \displaystyle\int (2x + 1)\, dx$

from which $\qquad\qquad y = x^2 + x + C$

This is the general solution of the given equation and represents the family of curves of which five typical ones are shown in Fig. 6.2.

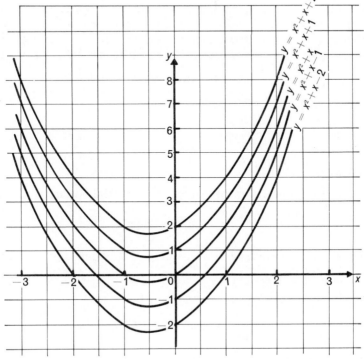

Fig. 6.2

To find the equation of the curve passing through the point $(3, 18)$ we must substitute these values of x and y in the general solution.

Hence
$$18 = 3^2 + 3 + C$$

∴
$$C = 6$$

and therefore the required equation is $y = x^2 + x + 6$.

BASIC DIFFERENTIAL EQUATIONS

Equations which contain a differential coefficient such as

$$\frac{dy}{dx} = 3 \quad \text{or} \quad \frac{dy}{dx} = 2x + 1$$

are called *differential equations*.

The expressions obtained by integration, for y in terms of x, and which include unknown constants, are called *general solutions*.

If a value of y corresponding to a value of x is known then this is called a *boundary condition*. Values given by a boundary condition may be substituted into a general solution and a numerical value obtained for the constant. The resulting equations, such as

$$y = 3x + 4 \quad \text{and} \quad y = x^2 + x + 6$$

are called *particular solutions*.

EXAMPLE 6.3

a) Find the general solution of the differential equation

$$\frac{dy}{dx} = 5x^3 + 2x - 7$$

b) From the boundary condition $y = 2$ when $x = 1$, find the particular solution.

a) We have
$$\frac{dy}{dx} = 5x^3 + 2x - 7$$

which may be rewritten as

$$y = \int (5x^3 + 2x - 7)\, dx$$

$$\therefore \qquad y = \tfrac{5}{4}x^4 + x^2 - 7x + C$$

This is the required general solution.

b) To find the particular solution we must substitute $y = 2$ when $x = 1$ into the general solution.

$$\therefore \qquad 2 = \tfrac{5}{4}(1)^4 + (1)^2 - 7(1) + C$$

from which
$$C = 2 - \tfrac{5}{4} - 1 + 7$$

$$= 6.75$$

Hence the required particular solution is

$$y = \tfrac{5}{4}x^4 + x^2 - 7x + 6.75$$

Exercise 6.1

1) Sketch the family of curves represented by the differential equation $\dfrac{dy}{dx} = x$, and find the equation of the curve passing through the point $(2, 3)$.

2) Sketch the family of curves represented by the differential equation $\dfrac{dy}{dx} = 3x^2$. Find also the equation of the curve which passes through the point $(5, -3)$.

3) Find the general solution of the differential equation

$$\frac{dy}{dx} = x^2 - 5x$$

and the particular solution if $x = 4$ when $y = 0$.

4) If $\dfrac{dy}{dx} = 6x^3 + 5x^2 + 7$, find the particular solution if it represents the equation of the curve which passes through the point $(2, 2)$.

5) If $\dfrac{dy}{dx} = ax$, where a is a constant, find the particular solution if $x = 0$ when $y = 0$, and also $x = 2$ when $y = 4$.

EQUATIONS OF THE TYPE $\dfrac{dy}{dx} = ky$

We will show now that $y = A e^{kx}$ is the general solution of the differential equation $\dfrac{dy}{dx} = ky$. In order to do this we must prove that $y = A e^{kx}$ satisfies the equation $\dfrac{dy}{dx} = ky$.

Now $$y = A e^{kx} \qquad\qquad [1]$$

and differentiating with respect to x we have

$$\frac{dy}{dx} = A k e^{kx} \qquad\qquad [2]$$

Now for the given differential equation

$$\text{the LHS} \ = \ \frac{dy}{dx}$$

$$= \ Ake^{kx} \quad \text{from equation [2]}$$

$$= \ k(Ae^{kx})$$

$$= \ ky \qquad \text{from equation [1]}$$

$$= \ \text{RHS}$$

Hence the differential equation $\dfrac{dy}{dx} = ky$

has a general solution $y = Ae^{kx}$ where A is a constant.

EXAMPLE 6.4

Given the differential equation $\dfrac{dy}{dx} = 6y$ find:

a) the general solution, and

b) the particular solution if $x = 0$ when $y = 3$.

a) Here the equation is of the form $\dfrac{dy}{dx} = ky$ where $k = 6$.

Hence the general solution is $y = Ae^{6x}$.

b) Substituting the values $x = 0$ and $y = 3$ into the general solution we have

$$3 \ = \ Ae^{6 \times 0}$$

from which $\qquad A \ = \ 3 \quad \text{since} \quad e^0 \ = \ 1$

Hence the particular solution is $y = 3e^{6x}$.

EXAMPLE 6.5

In a particular problem the rate of change of distance s with respect to time t is known to be proportional to the distance s. Express this as a differential equation and find:

a) the particular solution if $s = 2$ when $t = 1$, and $s = 6$ when $t = 1.5$,

b) the value of s when $t = 3$.

In mathematical notation the rate of change of s with respect to t is $\dfrac{ds}{dt}$.

Now we are given $\dfrac{ds}{dt}$ is proportional to s,

that is, $\qquad \dfrac{ds}{dt} = ks$, where k is a constant.

The general solution of this differential equation is $s = Ae^{kt}$ where both A and k are constants. The boundary conditions given in part a) will enable us to find the values of these two constants.

a) Now $s = 2$ when $t = 1$, and substituting these values into the general solution we have

$$2 = Ae^{k \times 1} \qquad\qquad [1]$$

Similarly, since $s = 6$ when $t = 1.5$

$$6 = Ae^{k \times 1.5} \qquad\qquad [2]$$

Dividing equation [2] by equation [1] gives

$$\frac{6}{2} = \frac{e^{1.5k}}{e^k}$$

$\therefore \qquad\qquad\qquad 3 = e^{(1.5-1)k}$

or $\qquad\qquad\qquad 3 = e^{0.5k}$

To solve this equation for k we must put it into logarithmic form,

that is $\qquad\qquad 0.5k = \log_e 3$

$\therefore \qquad\qquad\qquad k = \dfrac{1.10}{0.5}$

$\therefore \qquad\qquad\qquad k = 2.20$

To find A we will substitute $k = 2.20$ into equation [1]

$\therefore \qquad\qquad\qquad 2 = Ae^{2.20}$

or $\qquad\qquad\qquad A = \dfrac{2}{9.03}$

$\therefore \qquad\qquad\qquad A = 0.221$

Hence the required solution is $s = 0.221e^{2.20t}$

b) When $t = 3$ the corresponding value of s may be found by substituting $t = 3$ into the particular solution:

$$s = 0.221e^{2.20 \times 3}$$

$$= 162$$

EXAMPLE 6.6

Fig. 6.3 shows the general arrangement of a belt drive.

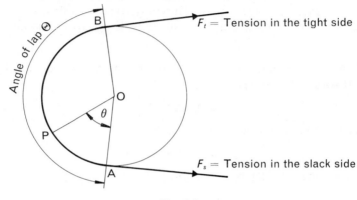

Fig. 6.3

It can be shown that the rate of change of the tension F N with respect to the angle θ radian at any point P is given by μF, where μ is the coefficient of friction between the belt and the pulley.

Find the equation connecting F_t, F_s, μ and Θ and hence find the value of F_t when $F_s = 50$ N, $\mu = 0.3$ and the angle of lap is $120°$.

The initial differential equation will be

$$\frac{\mathrm{d}F}{\mathrm{d}\theta} = \mu F$$

and hence the general solution will be of the form

$$F = Ae^{\mu\theta}$$

Now, when $\theta = 0$, $F = F_s$

$\therefore$
$$F_s = Ae^{\mu \times 0}$$

and since $e^0 = 1$

then
$$A = F_s$$

and hence the solution becomes

$$F = F_s e^{\mu\theta}$$

Also, when $\theta = \Theta$, $F = F_t$

$\therefore$
$$F_t = F_s e^{\mu\Theta'}$$

Hence the required equation is

$$F_t = F_s e^{\mu\Theta}$$

As in many mathematical equations the angle must be expressed in radians.

Since $360° = 2\pi$ radians

then $120° = \tfrac{2}{3}\pi = 2.09$ radians

Hence substituting the given values into the solution we have

$$F_t = 50e^{0.3 \times 2.09}$$

$$= 93.6 \text{ N}$$

EXAMPLE 6.7

Newton's law of cooling states that the rate at which the temperature T of a body falls is proportional to the difference in temperature between the body and its surroundings.

If t is the time and the surroundings are at $0°C$ then the differential equation representing the above information is $\dfrac{dT}{dt} = -kT$, the negative sign indicating a temperature drop.

If a body's temperature falls from $90°C$ to $70°C$ in 50 seconds find how long it will take to cool another $20°C$.

Since the differential equation given is

$$\frac{dT}{dt} = -kT$$

then the general solution is of the form

$$T = Ae^{-kt}$$

If we assume that cooling starts at $90°C$, then $T = 90$ when $t = 0$ s, and substituting these values into the general equation we have:

$$90 = Ae^{-k \times 0}$$

and since $e^0 = 1$, then $A = 90$

Also, when $T = 70\,^{\circ}$C then $t = 50\,$s, and substituting these values into $T = 90\mathrm{e}^{-kt}$ we get

$$70 = 90\mathrm{e}^{-k \times 50}$$

$$\therefore \qquad \mathrm{e}^{50k} = \frac{90}{70}$$

and rearranging into logarithmic form gives us

$$50k = \log_{\mathrm{e}}\left(\frac{90}{70}\right)$$

$$\therefore \qquad k = 0.005\,03$$

Hence the particular solution is

$$T = 90\mathrm{e}^{-0.005\,03t}$$

We shall now find the time to cool another $20\,^{\circ}$C, that is to $50\,^{\circ}$C. If we substitute $T = 50\,^{\circ}$C into $T = 90\mathrm{e}^{-0.005\,03t}$ the value of t will be the time to cool from $90\,^{\circ}$C, that is when the time commences, to $50\,^{\circ}$C.

Substituting gives us $\qquad 50 = 90\mathrm{e}^{-0.005\,03t}$

$$\mathrm{e}^{0.005\,03t} = \frac{90}{50}$$

and rearranging in logarithmic form

$$0.005\,03t = \log_{\mathrm{e}}\left(\frac{90}{50}\right)$$

$$t = 117\,\mathrm{s}$$

Hence the time to cool from $70\,^{\circ}$C to $50\,^{\circ}$C $= 117 - 50$

$$= 67\,\text{seconds}$$

Exercise 6.2

1) Find the general solution of the differential equation $\dfrac{\mathrm{d}y}{\mathrm{d}x} = 3y$ and the particular solution if $x = 0$ when $y = 2$. Find also the value of y when $x = 2$.

2) Find the particular solution of the differential equation $\dfrac{ds}{dt} = -ks$ if $s = 7$ when $t = 1$, and also $s = 4$ when $t = 2$. Find also the value of t when $s = 2$.

3) The rate at which the current I A dies in an electrical circuit which contains a resistance $R\,\Omega$ and an inductance L H is given by the differential equation $\dfrac{dI}{dt} = -\dfrac{R}{L}I$ where t s is time. If $R = 2\,\Omega$ and $L = 0.06$ H, and $I = 10$ A when $t = 0$ s, find the solution of the differential equation. Find also the current after 0.02 s.

4) A radioactive material decays at a rate which is proportional to the amount of radioactivity remaining. If the amount of radio-activity remaining is denoted by N at the time t, form a differential equation which represents this statement. If the half-life (that is the time taken for half of the radioactivity to decay) is 10 years, how long will it take for $\frac{3}{4}$ of the radioactivity to decay?

5) An electrical circuit contains a resistance R ohm and a capacitor C farad which initially holds a charge Q coulomb. The rate of discharge of the capacitor is given by the equation $\dfrac{dQ}{dt} + \dfrac{Q}{RC} = 0$, where t seconds is the time. If $R = 80\,000$ ohm and $C = 0.3 \times 10^{-6}$ farad, and also $Q = 0.0015$ coulomb initially, find the equation connecting Q and t. Find Q when $t = 0.02$ seconds and also t when $Q = 0.001$ coulomb.

7. NATURAL LOGARITHMS

On reaching the end of this chapter you should be able to:

1. *Define a natural (or Napierian) logarithm.*
2. *Determine natural logarithms from tables and by calculator.*
3. *State the relationship between natural and common logarithms.*
4. *Use natural logarithms to evaluate expressions arising from technology.*
5. *Solve equations involving e^x and $\log_e x$.*

THEORY OF LOGARITHMS

If N is a number such that

$$N = a^x$$

we say that x is the logarithm of N to the base a. We write

$$\log_a N = x$$

It should be carefully noted that

$$\text{NUMBER} = \text{BASE}^{\text{LOGARITHM}}$$

For example

since $\quad 100 = 10^2 \quad$ we may write $\quad \log_{10} 100 = 2$

and as

$$4.48 = e^{1.5} \quad \text{we may write} \quad \log_e 4.48 = 1.5$$

RULES FOR THE USE OF LOGARITHMS

These rules are true for any chosen value of the base:

(a) The logarithm of two numbers multiplied together may be found by adding their individual logarithms:

$$\log xy = \log x + \log y$$

(b) The logarithm of two numbers divided may be found by subtracting their individual logarithms:

$$\log \frac{x}{y} = \log x - \log y$$

(c) The logarithm of a number raised to a power may be found by multiplying the power by the logarithm of the number:

$$\log x^n = n \log x$$

LOGARITHMS TO THE BASE 10

Logarithms to the base 10 are called *common logarithms* and stated as $\log_{10}$ (or lg). When using logarithmic tables to solve numerical problems, we prefer tables to this base as they are simpler to use than tables to any other base. Common logarithms are also used for scales on logarithmic graph paper and also for calculations on the measurement of sound.

LOGARITHMS TO THE BASE e

In higher mathematics all logarithms are taken to the base e, where e = 2.718 28. Logarithms to this base are often called natural logarithms. They are also called Napierian or hyperbolic logarithms.

Natural logarithms are stated as $\log_e$ (or ln).

CHOICE OF BASE FOR CALCULATIONS

If a scientific calculator is used then it is just as easy to use logarithms to the base e. Some machines have keys for both $\log_e$ and $\log_{10}$ but on the more limited models only $\log_e$ is given.

The natural logarithms is found by using the $\log_e$ (or ln) key and the natural antilogarithm is found by using the e^x key.

The common logarithm is found using the $\log_{10}$ (or lg) key and the common antilogarithm is found using the 10^x key.

NATURAL LOGARITHMIC TABLES

In most books of mathematical tables there is a table of natural logarithms. Part of such a table is shown below:

Hyperbolic, Natural or Naperian Logarithms

4.5	1.5041	5063	5085	5107	5129	5151	5173	5195	5217	5239	2 4 7 9 11 13 15 18 20
4.6	1.5261	5282	5304	5326	5347	5369	5390	5412	5433	5454	2 4 6 9 11 13 15 17 19
4.7	1.5476	5497	5518	5539	5560	5581	5602	5623	5644	5665	2 4 6 8 11 13 15 17 19
4.8	1.5686	5707	5728	5748	5769	5790	5810	5831	5851	5872	2 4 6 8 10 12 14 16 19
4.9	1.5892	5913	5933	5953	5974	5994	6014	6034	6054	6074	2 4 6 8 10 12 14 16 18
5.0	1.6094	6114	6134	6154	6174	6194	6214	6233	6253	6273	2 4 6 8 10 12 14 16 18
5.1	1.6292	6312	6332	6351	6371	6390	6409	6429	6448	6467	2 4 6 8 10 12 14 16 18
5.2	1.6487	6506	6525	6544	6563	6582	6601	6620	6639	6658	2 4 6 8 10 11 13 15 17
5.3	1.6677	6696	6714	6734	6752	6771	6790	6808	6827	6845	2 4 6 7 9 11 13 15 17
5.4	1.6864	6882	6901	6919	6938	6956	6974	6993	7011	7029	2 4 5 7 9 11 13 15 17

Natural logarithms of 10^{+n}

n	1	2	3	4	5	6	7	8	9
$\log_e 10^n$	2.3026	4.6052	6.9078	9.2103	11.5129	13.8155	16.1181	18.4207	20.7233

The first column of a set of full tables gives the natural logarithms of numbers from 1.0 to 9.9 (the specific table above only gives numbers from 4.5 to 5.49), and the tables are read in the same way as ordinary log tables except that the characteristic is also given. Thus

$$\log_e 4.568 = 1.5191$$

When a natural logarithm of a number, which lies outside the tabulated range, is required the subsidiary table has to be used. The following examples show how this is done.

EXAMPLE 7.1

To find $\log_e 483.4$

Now
$$483.4 = 4.834 \times 100 = 4.834 \times 10^2$$

∴
$$\log_e 483.4 = \log_e (4.834 \times 10^2)$$
$$= \log_e 4.834 + \log_e 10^2$$

From the main table
$$\log_e 4.834 = 1.5756$$

From the subsidiary table
$$\log_e 10^2 = 4.6052$$

Thus
$$\log_e 483.4 = 1.5756 + 4.6052 = 6.1808$$

EXAMPLE 7.2

To find $\log_e 0.053\,61$

Now $\qquad\qquad 0.053\,61 = \dfrac{5.361}{100} = \dfrac{5.361}{10^2}$

$\therefore\qquad\qquad \log_e 0.053\,61 = \log_e\left(\dfrac{5.361}{10^2}\right)$

$\qquad\qquad\qquad\qquad = \log_e 5.361 - \log_e 10^2$

From the main table

$\qquad\qquad \log_e 5.361 = 1.6792$

From the subsidiary table

$\qquad\qquad \log_e 10^2 = 4.6052$

Thus $\qquad \log_e 0.053\,61 = 1.6792 - 4.6052$

$\qquad\qquad\qquad\qquad = -2.9260$

CONVERSION OF COMMON LOGARITHMS TO NATURAL LOGARITHMS

The use of tables of natural logarithms is rather tedious as the last two worked examples show. Their use may be avoided by finding the common logarithm and then converting to a natural logarithmic value.

Suppose we have a number, N, and wish to find the value of its natural logarithm, that is $\log_e N$.

We have: $\qquad \log_e N = \dfrac{\log_{10} N}{0.4343}$

or $\qquad\qquad \log_e N = 2.3026 \times \log_{10} N$

Hence we may find the natural logarithm of a number by first finding its common logarithm and then multiplying by 2.3026 (or dividing by 0.4343).

CALCULATIONS INVOLVING THE EXPONENTIAL FUNCTIONS e^x and $\log_e x$

EXAMPLE 7.3

Evaluate $50(e^{2.16})$.

The sequence of operations on a calculator is

| AC | 2 | . | 1 | 6 | e^x | × | 5 | 0 | = |

giving an answer 434 correct to 3 significant figures.

EXAMPLE 7.4

Evaluate $200(e^{-1.34})$.

The sequence of operations would then be

| AC | 1 | . | 3 | 4 | +/− | e^x |

| × | 2 | 0 | 0 | = |

giving an answer 52.4 correct to 3 significant figures.

EXAMPLE 7.5

In a capacitive circuit the instantaneous voltage across the capacitor is given by $v = V(1 - e^{-t/CR})$ where V is the initial supply voltage, R ohms the resistance, C farads the capacitance, and t seconds the time from the instant of connecting the supply voltage.

If $V = 200$, $R = 10\,000$, and $C = 20 \times 10^{-6}$ find the time when the voltage v is 100 volts.

Substituting the given values in the equation we have

$$100 = 200(1 - e^{-t/20 \times 10^{-6} \times 10000})$$

$$\therefore \quad \frac{100}{200} = 1 - e^{-t/0.2}$$

$$\therefore \quad 0.5 = 1 - e^{-5t}$$

$$\therefore \quad e^{-5t} = 1 - 0.5$$

$$\therefore \quad e^{-5t} = 0.5$$

Thus in log form

$$\log_e 0.5 = -5t$$

$$\therefore \qquad t = -\frac{\log_e 0.5}{5}$$

The sequence of operation is

| AC | 0 | . | 5 | ln | ÷ | 5 | = | +/− |

giving an answer 0.139 seconds correct to 3 significant figures.

EXAMPLE 7.6

$$R = \frac{(0.42)S}{l} \times \log_e\left(\frac{d_2}{d_1}\right)$$

refers to the insulation resistance of a wire. Find the value of R when $S = 2000$, $l = 120$, $d_1 = 0.2$ and $d_2 = 0.3$

Substituting the given values gives

$$R = \frac{0.42 \times 2000}{120} \times \log_e\left(\frac{0.3}{0.2}\right)$$

$$= \frac{0.42 \times 2000}{120} \times \log_e 1.5$$

The sequence of operations would be

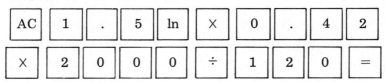

| AC | 1 | . | 5 | ln | × | 0 | . | 4 | 2 |
| × | 2 | 0 | 0 | 0 | ÷ | 1 | 2 | 0 | = |

giving an answer 2.84 correct to 3 significant figures.

Exercise 7.1

1) Find the numbers whose natural logarithms are:

(a) 2.76   (b) 0.677   (c) 0.09   (d) -3.46

(e) -0.543 (f) -0.078

2) Find the values of:

(a) $70e^{2.5}$ (b) $150e^{-1.34}$ (c) $3.4e^{-0.445}$

3) The formula

$$L = 0.000\,644 \left\{ \log_e \frac{d}{r} + \frac{1}{4} \right\}$$

is used for calculating the self-inductance of parallel conductors. Find L when $d = 50$ and $r = 0.25$

4) The inductance (L microhenrys) of a straight aerial is given by the formula

$$L = \frac{1}{500} \left(\log_e \frac{4l}{d} - 1 \right)$$

where l is the length of the aerial in mm and d its diameter in mm. Calculate the inductance of an aerial 5000 mm long and 2 mm in diameter.

5) Find the value of $\log_e \left(\frac{c_1}{c_2} \right)^2$ when $c_1 = 4.7$ and $c_2 = 3.5$

6) If $T = R \log_e \left(\frac{a}{a-b} \right)$ find T when $R = 28$, $a = 5$ and $b = 3$.

7) When a chain of length $2l$ is suspended from two points $2d$ apart on the same horizontal level

$$d = c \log_e \left(\frac{l + \sqrt{l^2 + c^2}}{c} \right)$$

If $c = 80$ and $l = 200$, find d.

8) The instantaneous value of the current when an inductive circuit is discharging is given by the formula $i = e^{-Rt/L}$. Find the value of this current, i, when $R = 30$, $L = 0.5$ and $t = 0.005$

9) In a circuit in which a resistor is connected in series with a capacitor the instantaneous voltage across the capacitor is given by the formula $v = V(1 - e^{-t/CR})$. Find this voltage, v, when $V = 200$, $C = 40 \times 10^{-6}$, $R = 100\,000$ and $t = 1$.

10) In the formula $v = Ve^{-Rt/L}$ the values of v, V, R and L are 50, 150, 60 and 0.3 respectively. Find the corresponding value of t.

11) The instantaneous charge in a capacitive circuit is given by

$$q = Q(1 - e^{-t/CR})$$

Find the value of t when $q = 0.01$, $Q = 0.015$, $C = 0.0001$, and $R = 7000$.

PARTIAL FRACTIONS

COMPOUND FRACTIONS AND PARTIAL FRACTIONS

Suppose we wish to express $\dfrac{2}{(x+1)} + \dfrac{3x+6}{(x^2+2)}$ as a single fraction.

The LCM of the denominator is $(x+1)(x^2+2)$ and so

$$\frac{2}{(x+1)} + \frac{3x+6}{(x^2+2)} = \frac{2(x^2+2)+(3x+6)(x+1)}{(x+1)(x^2+2)}$$

$$= \frac{2x^2+4+3x^2+3x+6x+6}{(x+1)(x^2+2)} = \frac{5x^2+9x+10}{(x+1)(x^2+2)}$$

It is often necessary to reverse the above procedure and express

$$\frac{5x^2+9x+10}{(x+1)(x^2+2)} \quad \text{as} \quad \frac{2}{(x+1)} + \frac{3x+6}{(x^2+2)}$$

Now $\dfrac{5x^2+9x+10}{(x+1)(x^2+2)}$ is known as a compound fraction while

$\dfrac{2}{(x+1)}$ and $\dfrac{3x+6}{(x^2+2)}$ are known as partial fractions.

We see that in each of these fractions the numerator is of a smaller degree than the denominator. Remember that the degree of an expression depends on the highest power of the variable — thus in the second partial fraction the numerator $3x+6$ contains the first power of x and is of the first degree while the denominator x^2+2 contains the second power of x and is of the second degree.

HOW TO FIND PARTIAL FRACTIONS

EXAMPLE 8.1

Express $\dfrac{x+2}{(1+x)(2-x)}$ in terms of partial fractions.

The numerator $x+2$ is of the first degree while the denominator $(1+x)(2-x) = 2+x-x^2$ is of the second degree and thus we shall be able to find suitable partial fractions.

Now only experience will enable you to decide the form of the partial fractions, but we must remember that in each case the numerator must be of a smaller degree than the denominator. If A and B are constants (i.e. numbers), then let

$$\frac{x+2}{(1+x)(2-x)} \equiv \frac{A}{(1+x)} + \frac{B}{(2-x)}$$

or

$$\frac{x+2}{(1+x)(2-x)} \equiv \frac{A(2-x)+B(1+x)}{(1+x)(2-x)}$$

You will note that the indentity sign is used which means that the expression is true for any value of the variable, x. Comparing both sides of the identity we see that the denominators are the same and so we must arrange for the numerators to be the same also.

Thus
$$x+2 = A(2-x)+B(1+x)$$

Now by choosing suitable values of x it is possible to find the values of the constants A and B.

When $x = 2$
$$2+2 = A(2-2)+B(1+2)$$
$$\therefore \qquad 4 = A\times0 + B\times3$$

from which
$$B = \tfrac{4}{3}$$

also when $x = -1$
$$-1+2 = A[2-(-1)] + B[1+(-1)]$$
$$\therefore \qquad 1 = A\times3 + B\times0$$

from which
$$A = \tfrac{1}{3}$$

Thus
$$\frac{x+2}{(1+x)(2-x)} \equiv \frac{\tfrac{1}{3}}{(1+x)} + \frac{\tfrac{4}{3}}{(2-x)} \equiv \frac{1}{3(1+x)} + \frac{4}{3(2-x)}$$

EXAMPLE 8.2

Express $\dfrac{2x^2 + 7x - 17}{(x-1)(x-2)(x+3)}$ in terms of partial fractions.

This is possible since the numerator is of the second degree, which is one less than the degree of the denominator.

Let $\dfrac{2x^2 + 7x - 17}{(x-1)(x-2)(x+3)} \equiv \dfrac{A}{(x-1)} + \dfrac{B}{(x-2)} + \dfrac{C}{(x+3)}$

or $\dfrac{2x^2 + 7x - 17}{(x-1)(x-2)(x+3)} \equiv \dfrac{\begin{array}{c}A(x-2)(x+3) + B(x-1)(x+3) \\ + C(x-1)(x-2)\end{array}}{(x-1)(x-2)(x+3)}$

The denominators are the same, and we must arrange that the numerators are identical.

Thus $2x^2 + 7x - 17 \equiv A(x-2)(x+3) + B(x-1)(x+3)$
$\qquad\qquad\qquad\qquad\qquad\quad + C(x-1)(x-2)$

When $x = 1$

$2(1)^2 + 7(1) - 17 = A(1-2)(1+3) + B(1-1)(1+3)$
$\qquad\qquad\qquad\qquad\qquad\quad + C(1-1)(1-2)$

$\therefore \qquad\qquad 2 + 7 - 17 = A(-1)(4) + B(0)(4) + C(0)(-1)$

from which $\qquad\qquad A = 2$

and when $x = -3$

$2(-3)^2 + 7(-3) - 17 = \quad A(-3-2)(-3+3)$
$\qquad\qquad\qquad\qquad\qquad + B(-3-1)(-3+3)$
$\qquad\qquad\qquad\qquad\qquad + C(-3-1)(-3-2)$

$\therefore \qquad\qquad 18 - 21 - 17 = A(-5)(0) + B(-4)(0) + C(-4)(-5)$

from which $\qquad\qquad C = -1$

and when $x = 2$

$2(2)^2 + 7(2) - 17 = A(2-2)(2+3) + B(2-1)(2+3)$
$\qquad\qquad\qquad\qquad\qquad + C(2-1)(2-2)$

$\therefore \qquad\qquad 8 + 14 - 17 = A(0)(5) + B(1)(5) + C(1)(0)$

from which $\qquad\qquad B = 1$

Thus

$\dfrac{2x^2 + 7x - 17}{(x-1)(x-2)(x+3)} \equiv \dfrac{2}{(x-1)} + \dfrac{1}{(x-2)} - \dfrac{1}{(x+3)}$

EXAMPLE 8.3

Express $\dfrac{6x^2 + 9x + 1}{(x + 1)^2(x - 1)}$ in terms of partial fractions.

This is possible since the numerator is of the second degree which is smaller than the denominator which is the third degree. Now when the denominator of the given compound fraction contains a factor to a power – here $(x + 1)^2$ – then the partial fractions must be of the form shown.

Let $\dfrac{6x^2 + 9x + 1}{(x + 1)^2(x - 1)} \equiv \dfrac{A}{(x + 1)^2} + \dfrac{B}{(x + 1)} + \dfrac{C}{(x - 1)}$

or $\dfrac{6x^2 + 9x + 1}{(x + 1)^2(x - 1)} \equiv \dfrac{A(x - 1) + B(x + 1)(x - 1) + C(x + 1)^2}{(x + 1)^2(x - 1)}$

The denominators are the same, and we must arrange for the numerators to be identical.

Thus $\quad 6x^2 + 9x + 1 \equiv A(x - 1) + B(x + 1)(x - 1) + C(x + 1)^2$

When $x = 1$

$$6(1)^2 + 9(1) + 1 = A(1 - 1) + B(1 + 1)(1 - 1) + C(1 + 1)^2$$

$\therefore \qquad 6 + 9 + 1 = A(0) + B(2)(0) + C(2)^2$

from which $\qquad C = 4$

and when $x = -1$

$$6(-1)^2 + 9(-1) + 1 = A(-1 - 1) + B(-1 + 1)(-1 - 1)$$
$$+ C(-1 + 1)^2$$

$\therefore \qquad 6 - 9 + 1 = A(-2) + B(0)(-2) + C(0)^2$

from which $\qquad A = 1$

We now use the facts that $A = 1$ and $C = 4$ and choose any value for x other than 1 or -1 used previously. For simplicity we will use $x = 0$.

Then

$$6(0)^2 + 9(0) + 1 = 1(0 - 1) + B(0 + 1)(0 - 1) + 4(0 + 1)^2$$

from which $\qquad B = 2$

Thus

$$\dfrac{6x^2 + 9x + 1}{(x + 1)^2(x - 1)} \equiv \dfrac{1}{(x + 1)^2} + \dfrac{2}{(x + 1)} + \dfrac{4}{(x - 1)}$$

EXAMPLE 8.4

Express $\dfrac{3x^2 + 5x + 1}{(x^2 + 4)(x + 3)}$ in terms of partial fractions.

Again the problem is possible since the numerator is of a lesser degree than the denominator. Here again the form of the partial fractions is known by experience and you should remember this.

Let

$$\frac{3x^2 + 5x + 1}{(x^2 + 4)(x + 3)} = \frac{Ax + B}{(x^2 + 4)} + \frac{C}{(x + 3)}$$

or

$$\frac{3x^2 + 5x + 1}{(x^2 + 4)(x + 3)} = \frac{(Ax + B)(x + 3) + C(x^2 + 4)}{(x^2 + 4)(x + 3)}$$

The denominators are the same and so we must equate the numerators.

Thus $\quad 3x^2 + 5x + 1 \equiv (Ax + B)(x + 3) + C(x^2 + 4)$

When $x = -3$

$$3(-3)^2 + 5(-3) + 1 = [A(-3) + B](-3 + 3) \\ + C[(-3)^2 + 4]$$

$$\therefore \qquad\qquad 27 - 15 + 1 = (-3A + B)(0) + C(9 + 4)$$

from which $\qquad\qquad C = 1$

It is not possible to find the values of the remaining constants A and B by substituting specific values of x —try it for yourself. An alternative method is used in which we equate 'like' terms on the left and right-hand sides of the identity. This will necessitate multiplying out the right-hand side.

Hence $\qquad 3x^2 + 5x + 1 \equiv (Ax + B)(x + 3) + C(x^2 + 4)$

will become $\quad 3x^2 + 5x + 1 \equiv Ax^2 + 3Ax + Bx + 3B + Cx^2 + 4C$

and putting $C = 1$ gives

$$3x^2 + 5x + 1 = (A + 1)x^2 + (3A + B)x + (3B + 4)$$

Equating coefficients of x^2 gives

$$3 = A + 1$$

$$\therefore \qquad\qquad A = 2$$

and equating coefficients of x gives

$$5 = (3A + B)$$

But $A = 2$, thus $5 = (3 \times 2) + B$

from which $B = -1$

Thus $\dfrac{3x^2 + 5x + 1}{(x^2 + 4)(x + 3)} \equiv \dfrac{2x - 1}{(x^2 + 4)} + \dfrac{1}{(x + 3)}$

EXAMPLE 8.5

Express $\dfrac{x^2 + 5x + 5}{(x + 1)(x + 2)}$ in terms of partial fractions.

We cannot proceed here as the degree of the numerator is the same as that of the denominator. We must, therefore, divide the denominator into the numerator. The denominator $(x + 1)(x + 2)$ when multiplied out is $x^2 + 3x + 2$.

Thus $x^2 + 3x + 2) \; x^2 + 5x + 5 \; (\; 1$

$$x^2 + 3x + 2$$

$$\overline{}$$

$$2x + 3$$

Hence the result is 1 together with a remainder of $2x + 3$

or $\dfrac{x^2 + 5x + 5}{(x + 1)(x + 2)} \equiv 1 + \dfrac{2x + 3}{(x + 1)(x + 2)}$

The compound fraction on the right-hand side of the identity has a numerator of lesser degree than the denominator and so may be expressed in terms of partial fractions in a manner similar to that used in Example 8.1. You may care to check this yourself.

Hence $\dfrac{x^2 + 5x + 5}{(x + 1)(x + 2)} \equiv 1 + \dfrac{1}{(x + 1)} + \dfrac{1}{(x + 2)}$

Exercise 8.1

Express in terms of partial fractions:

1) $\dfrac{5x - 3}{(x - 3)(x + 3)}$ 2) $\dfrac{x}{(x - 1)(x - 2)}$

3) $\dfrac{5x - 7}{(x + 1)(2x - 1)}$ 4) $\dfrac{x + 7}{x^2 + 3x + 2}$

5) $\dfrac{9x^2 + 34x + 29}{(x+1)(x+2)(x+3)}$

6) $\dfrac{3x^2 - 7x + 2}{(2x-1)(x+1)(x-1)}$

7) $\dfrac{6 - 15x + 7x^2}{2x(1-x)(2-3x)}$

8) $\dfrac{2x+4}{(1+x)(x-1)(2x+1)}$

9) $\dfrac{7x - 2x^2}{(2-x)^2(1+x)}$

10) $\dfrac{10 - 2x - 3x^2}{(x-1)^2(2x+3)}$

11) $\dfrac{x^2 + 3}{(x+1)(x-1)^2}$

12) $\dfrac{-(2+3x)}{x(1+x)^2}$

13) $\dfrac{3x^2 + 3x + 2}{(x^2+1)(x+1)}$

14) $\dfrac{2x^2 + 2x - 7}{(x-1)(2+x^2)}$

15) $\dfrac{10 + 5x - 8x^2}{(2-3x^2)(4+x)}$

16) $\dfrac{6x^2 + 8x + 16}{(x+2)(x^2+4)}$

17) $\dfrac{x^2 + 4x - 2}{(x-1)(x+2)}$

18) $\dfrac{x(2x-3)}{(2x-1)(x-1)}$

19) $\dfrac{x^2 - 3x - 2}{(x-1)(x+1)}$

20) $\dfrac{3(x^2 - x - 1)}{(x+1)(x-2)}$

THE BINOMIAL THEOREM

On reaching the end of this chapter you should be able to:

1. *Expand expressions of the form* $(a + x)^n$ *for small, positive integers n.*
2. *State the general form for the binomial coefficients for all positive integers n.*
3. *Expand expressions of the form* $(1 + x)^n$ *where n takes positive, negative or fractional values.*
4. *State the range of values of x for which the series is convergent.*
5. *Calculate the effect on the subject of a formula when one or more of the independent variables is subject to a small change or error.*

BINOMIAL EXPRESSION

A *binomial expression* consists of two terms. Thus $1 + x$, $a + b$, $5y - 2$, $3x^2 + 7$ and $7a^3 + 3b^2$ are all binomial expressions. The *binomial theorem* allows us to expand powers of such expressions.

THE BINOMIAL THEOREM

Now $(a + b)^0 = 1$

(since any number to the power 0 is unity),

and also $(a + b)^1 = a + b$

Multiplying both sides by $(a + b)$ gives

$$(a + b)^2 = a^2 + 2ab + b^2$$

also $(a + b)^3 = a^3 + 3a^2b + 3ab^2 + b^3$

and $(a + b)^4 = a^4 + 4a^3b + 6a^2b^2 + 4ab^3 + b^4$

We can now arrange the coefficients of each of the terms of the above expansions in the form known as *Pascal's triangle*.

Binomial expression	Coefficients in the expansion

$(a+b)^0$

$\qquad\qquad$ 1

$(a+b)^1$

$\qquad$ 1 $\quad$ 1

$(a+b)^2$

1 $\quad$ 2 $\quad$ 1

$(a+b)^3$

1 $\quad$ 3 $\quad$ 3 $\quad$ 1

$(a+b)^4$

1 $\quad$ 4 $\quad$ 6 $\quad$ 4 $\quad$ 1

$(a+b)^5$

1 $\quad$ 5 $\quad$ 10 $\quad$ 10 $\quad$ 5 $\quad$ 1

$(a+b)^6$

1 $\quad$ 6 $\quad$ 15 $\quad$ 20 $\quad$ 15 $\quad$ 6 $\quad$ 1

$(a+b)^7$

1 $\quad$ 7 $\quad$ 21 $\quad$ 35 $\quad$ 35 $\quad$ 21 $\quad$ 7 $\quad$ 1

It will be seen that:

(a) The number of terms in each expansion is one more than the index. Thus the expansion of $(a+b)^9$ will have 10 terms.

(b) The arrangement of the coefficients is symmetrical.

(c) The coefficients of the first and last terms are both always unity.

(d) Each coefficient in the table is obtained by adding together the two coefficients in the line above, which lie on either side of it.

The expansion of $(a+b)^8$ is therefore:

$$a^8 + (1+7)a^7b + (7+21)a^6b^2 + (21+35)a^5b^3 + (35+35)a^4b^4$$
$$+ (35+21)a^3b^5 + (21+7)a^2b^6 + (7+1)ab^7 + b^8$$
$$= a^8 + 8a^7b + 28a^6b^2 + 56a^5b^3 + 70a^4b^4 + 56a^3b^5 + 28a^2b^6$$
$$+ 8ab^7 + b^8$$

It is inconvenient to use Pascal's triangle when expanding higher powers of $(a+b)$. In such cases the following series is used:

$$(a+b)^n = a^n + na^{n-1}b + \frac{n(n-1)}{2!}a^{n-2}b^2 + \frac{n(n-1)(n-2)}{3!}a^{n-3}b^3$$
$$+ \ldots + b^n$$

This is the *binomial theorem* and is true for all positive whole numbers n.

The symbol '!' indicates 'factorial' when following a positive whole number.

For example, 4! is pronounced 'factorial four' and means $4 \times 3 \times 2 \times 1$.

Thus $\quad 2! = 2 \times 1 \qquad\qquad 3! = 3 \times 2 \times 1$

$\qquad\quad 4! = 4 \times 3 \times 2 \times 1 \qquad 5! = 5 \times 4 \times 3 \times 2 \times 1$

$\qquad$ and so on.

EXAMPLE 9.1

Expand $(3x + 2y)^4$.

Comparing $(3x + 2y)^4$ with $(a + b)^4$, we have $3x$ in place of a, and $2y$ in place of b. Substituting in the standard expansion, we get

$$[(3x) + (2y)]^4 = (3x)^4 + 4(3x)^3(2y) + \frac{(4)(4-1)}{2!}(3x)^2(2y)^2$$

$$+ \frac{(4)(4-1)(4-2)}{3!}(3x)(2y)^3 + (2y)^4$$

$$= (3x)^4 + 4(3x)^3 2y + \frac{4 \times 3}{2 \times 1}(3x)^2(2y)^2$$

$$+ \frac{4 \times 3 \times 2}{3 \times 2 \times 1}(3x)(2y)^3 + (2y)^4$$

$$= 81x^4 + 216x^3 y + 216x^2 y^2 + 96xy^3 + 16y^4$$

EXAMPLE 9.2

Expand $(x - 4y)^{15}$ to four terms.

The given binomial expression should be rewritten as $[x + (-4y)]^{15}$ and comparing this with the standard expression for $(a + b)^n$ we have x in place of a, $-4y$ in place of b, and 15 in place of n.

$$\therefore \ [x + (-4y)]^{15} = x^{15} + 15x^{14}(-4y) + \frac{(15)(15-1)}{2!}x^{13}(-4y)^2$$

$$+ \frac{15(15-1)(15-2)}{3!}x^{12}(-4y)^3 + \ldots$$

$$= x^{15} + 15(-4)x^{14}y + \frac{15 \times 14 \times (-4)^2}{2 \times 1}x^{13}y^2$$

$$+ \frac{15 \times 14 \times 13 \times (-4)^3}{3 \times 2 \times 1}x^{12}y^3 + \ldots$$

$$= x^{15} - 60x^{14}y + 1680x^{13}y^2 - 29\,120x^{12}y^3 + \ldots$$

Exercise 9.1

Expand using Pascal's triangle:

1) $(1 + z)^5$ 2) $(p + q)^6$ 3) $(x - 3y)^4$

4) $(2p - q)^5$ 5) $(2x + y)^7$ 6) $\left(x + \dfrac{1}{x}\right)^3$

Expand to four terms using the binomial theorem:

7) $(1+x)^{12}$ 8) $(1-2x)^{14}$ 9) $(p+q)^{16}$

10) $(1+3y)^{10}$ 11) $(x^2-3y)^9$ 12) $\left(x^2+\dfrac{1}{x^2}\right)^{11}$

THE BINOMIAL SERIES

If we put $a = 1$ and use x instead of b in the binomial theorem we get

$$(1+x)^n = 1 + nx + \frac{n(n-1)}{2!}x^2 + \frac{n(n-1)(n-2)}{3!}x^3 + \ldots + x^n$$

This expression is true for all positive whole number values of n.

For negative and fractional values of n the right-hand side of the above expression has no final term. It has an infinite number of terms and is called an infinite series and forms the binomial series:

$$(1+x)^n = 1 + nx + \frac{n(n-1)}{2!}x^2 + \frac{n(n-1)(n-2)}{3!}x^3 + \ldots$$

There is also a proviso that, for negative and fractional values of n, any value given to x must lie between $+1$ and -1. The reason for this is explained under the heading 'Convergence of Series' on p. 101.

EXAMPLE 9.3

Use the binomial series to expand $\sqrt{(1+x)}$ to five terms.

$$\sqrt{(1+x)} = (1+x)^{1/2}$$

$$= 1 + \tfrac{1}{2}x + \frac{\tfrac{1}{2}(\tfrac{1}{2}-1)}{2!}x^2 + \frac{\tfrac{1}{2}(\tfrac{1}{2}-1)(\tfrac{1}{2}-2)}{3!}x^3$$

$$+ \frac{\tfrac{1}{2}(\tfrac{1}{2}-1)(\tfrac{1}{2}-2)(\tfrac{1}{2}-3)}{4!}x^4 + \ldots$$

$$= 1 + \tfrac{1}{2}x + \frac{(\tfrac{1}{2})(-\tfrac{1}{2})}{2\times1}x^2 + \frac{\tfrac{1}{2}(-\tfrac{1}{2})(-\tfrac{3}{2})}{3\times2\times1}x^3$$

$$+ \frac{\tfrac{1}{2}(-\tfrac{1}{2})(-\tfrac{3}{2})(-\tfrac{5}{2})}{4\times3\times2\times1}x^4 + \ldots$$

$$= 1 + \tfrac{1}{2}x - \frac{1}{2 \times 2 \times 2}x^2 + \frac{3}{2 \times 2 \times 2 \times 3 \times 2 \times 1}x^3$$

$$- \frac{3 \times 5}{2 \times 2 \times 2 \times 2 \times 4 \times 3 \times 2 \times 1}x^4 + \dots$$

$$= 1 + \tfrac{1}{2}x - \tfrac{1}{8}x^2 + \tfrac{1}{16}x^3 - \tfrac{5}{128}x^4 + \dots$$

EXAMPLE 9.4

Use the binomial series to expand $\dfrac{1}{(1 + 2x)^5}$ to four terms.

$$\frac{1}{(1 + 2x)^5} = [1 + (2x)]^{-5}$$

$$= 1 + (-5)(2x) + \frac{(-5)(-5-1)}{2!}(2x)^2$$

$$+ \frac{(-5)(-5-1)(-5-2)}{3!}(2x)^3 + \dots$$

$$= 1 - 10x + \frac{5 \times 6}{2 \times 1} \times 4x^2 - \frac{5 \times 6 \times 7}{3 \times 2 \times 1} \times 8x^3 + \dots$$

$$= 1 - 10x + 60x^2 - 280x^3 + \dots$$

EXAMPLE 9.5

Use the binomial series to expand $\dfrac{1}{\sqrt[4]{(1-x)}}$ to four terms.

$$\frac{1}{\sqrt[4]{(1-x)}} = [1 - x]^{-1/4}$$

$$= [1 + (-x)]^{-1/4}$$

$$= 1 + (-\tfrac{1}{4})(-x) + \frac{(-\tfrac{1}{4})(-\tfrac{1}{4}-1)}{2!}(-x)^2$$

$$+ \frac{(-\tfrac{1}{4})(-\tfrac{1}{4}-1)(-\tfrac{1}{4}-2)}{3!}(-x)^3 + \dots$$

$$= 1 + \tfrac{1}{4}x + \frac{(-\tfrac{1}{4})(-\tfrac{5}{4})}{2 \times 1}x^2 - \frac{(-\tfrac{1}{4})(-\tfrac{5}{4})(-\tfrac{9}{4})}{3 \times 2 \times 1}x^3 + \dots$$

$$= 1 + \tfrac{1}{4}x + \frac{5}{4 \times 4 \times 2}x^2 + \frac{5 \times 9}{4 \times 4 \times 4 \times 3 \times 2}x^3 + \ldots$$

$$= 1 + \tfrac{1}{4}x + \tfrac{5}{32}x^2 + \tfrac{15}{128}x^3 + \ldots$$

CONVERGENCE OF SERIES

Consider the binomial series obtained in Example 9.3 which gave

$$(1 + x)^{1/2} = 1 + \tfrac{1}{2}x - \tfrac{1}{8}x^2 + \tfrac{1}{16}x^3 - \tfrac{5}{128}x^4 + \ldots$$

Putting $x = \tfrac{1}{2}$ we get

$$(1 + \tfrac{1}{2})^{1/2} = 1 + \tfrac{1}{2}(\tfrac{1}{2}) - \tfrac{1}{8}(\tfrac{1}{2})^2 + \tfrac{1}{16}(\tfrac{1}{2})^3 - \tfrac{5}{128}(\tfrac{1}{2})^4 + \ldots$$

$$= 1 + 0.25 - 0.031\,25 + 0.007\,813 - 0.002\,441 + \ldots$$

$$= 1.22 \quad \text{correct to 3 significant figures}$$

We can see that the terms in the expansion when $x = \tfrac{1}{2}$ are rapidly getting smaller and a numerical answer to any reasonable degree of accuracy may soon be obtained.

Such a series is said to be 'convergent'.

Let us now put $x = 2$ into the same expansion, getting

$$(1 + 2)^{1/2} = 1 + \tfrac{1}{2}(2) - \tfrac{1}{8}(2)^2 + \tfrac{1}{16}(2)^3 - \tfrac{5}{128}(2)^4 + \ldots$$

$$= 1 + 1 - 0.5 + 0.5 - 0.625 + \ldots$$

In this series the terms do not get progressively smaller — in fact they subsequently become larger. This means it is not possible to use the series to find a numerical value and the series is *not* convergent.

In general when using the *binomial series*:

If index n is negative or fractional, then the value of x must lie between $+1$ and -1.

However if index n is a positive whole number the series is finite and a numerical answer may always be obtained.

Exercise 9.2

Expand to four terms using the binomial series:

1) $(1 + x)^6$

2) $(1 + 2x)^9$

3) $(1 + x)^{-4}$

4) $(1 - x)^{-1/2}$

5) $(1 + x)^{-1}$

6) $(1 + 2x)^{-3}$

7) $(1-x)^{-3}$ 8) $\sqrt[3]{(1+x^2)}$ 9) $\dfrac{1}{\sqrt{(1-3x)}}$

10) $\left(1-\dfrac{2x}{3}\right)^{3/4}$

APPLICATION TO SMALL ERRORS

The binomial series is

$$(1+x)^n = 1 + nx + \frac{n(n-1)}{2!}x^2 + \frac{n(n-1)(n-2)}{3!}x^3 + \ldots$$

When x is small when compared with unity, e.g. 0.03, then

$$(1+x)^n = 1 + nx \quad \text{approximately}$$

as all other terms in the series contain powers of x which are negligible when compared with the first two shown. This relationship is often useful.

It is easy enough to measure the expansion of a metal rod which has been heated. It is, however, not so easy to measure the increases in areas and volumes which have been expanded.

If α is the coefficient of linear expansion, then the length of an expanded bar originally $1\,m$ long is $(1+\alpha)\,m$ for one degree temperature increase. An area of the same material originally $1\,m \times 1\,m$ becomes $(1+\alpha)^2\,m^2$ when it expands. From the above approximation as α is very small then

$$(1+\alpha)^2 \simeq 1 + 2\alpha$$

Hence *the area coefficient is approximately twice the linear coefficient.*

Similarly a volume $1\,m \times 1\,m \times 1\,m$ on expanding becomes $(1+\alpha)^3\,m^3$ and again, using the above approximation:

$$(1+\alpha)^3 \simeq 1 + 3\alpha$$

which shows that *the volume coefficient is approximately three times the linear coefficient.*

EXAMPLE 9.6

In measuring the radius of a circle, the measurement is 1% too large. If this measurement is used to calculate the area of the circle find the resulting error.

Let A be the area of the circle, and r the radius. If δA is the error in the area, then

$$A = \pi r^2$$

and
$$A + \delta A = \pi(r + r \times \tfrac{1}{100})^2$$

$$= \pi r^2(1 + \tfrac{1}{100})^2$$

and since $\tfrac{1}{100}$ is small when compared with 1, then

$$A + \delta A \simeq \pi r^2(1 + 2 \times \tfrac{1}{100})$$

$$\simeq A(1 + \tfrac{2}{100})$$

$$\delta A \simeq \tfrac{2}{100} A$$

Hence
$$\delta A \simeq 2\% \text{ of } A$$

Thus an error of 1% too large in the radius gives an approximate error of 2% too large in the area.

EXAMPLE 9.7

When a uniform beam is simply supported at its ends the deflection at the centre of span is given by

$$y = \frac{5wl^4}{384EI}$$

where w is the distributed load per unit length, l is the length between supports, E is Young's modulus and I is the 2nd moment of cross-sectional area. Find the percentage change in y when l decreases by 3%.

Let the change in y be δy. Then

$$y + \delta y = \frac{5w}{384EI}[l - l \times \tfrac{3}{100}]^4$$

$$= \frac{5wl^4}{384EI}[1 + (-\tfrac{3}{100})]^4$$

$$= \frac{5wl^4}{384EI}[1 + 4(-\tfrac{3}{100})] \quad \text{approximately}$$

since all higher powers of $\tfrac{3}{100}$ are negligible.

Thus
$$y + \delta y \simeq y[1 - \tfrac{12}{100}]$$

$$\therefore \qquad \delta y \simeq -\tfrac{12}{100}y \quad \text{or} \quad 12\% \text{ of } y \text{ (decrease)}$$

Hence y will decrease by 12% approximately.

EXAMPLE 9.8

A formula used in connection with close-coiled helical springs is

$$P = \frac{GFd^5}{8hD^3}$$

Find the change in P if d is increased by 2% and D is decreased by 3%.

Let δP be the change in P. Then

$$
\begin{aligned}
P + \delta P &= \frac{GF[d + \frac{2}{100}d]^5}{8h[D - \frac{3}{100}D]^3} \\
&= \frac{GFd^5[1 + \frac{2}{100}]^5}{8hD^3[1 - \frac{3}{100}]^3} \\
&= P[1 + (\tfrac{2}{100})]^5[1 + (-\tfrac{3}{100})]^{-3} \\
&\simeq P[1 + 5(\tfrac{2}{100})][1 + (-3)(-\tfrac{3}{100})]
\end{aligned}
$$

since all higher powers of $\frac{2}{100}$ and $\frac{3}{100}$ are negligible.

Thus
$$
\begin{aligned}
P + \delta P &\simeq P[1 + \tfrac{10}{100}][1 + \tfrac{9}{100}] \\
&= P[1 + \tfrac{19}{100}] \quad \text{approximately} \\
&= P + \tfrac{19}{100}P
\end{aligned}
$$

$\therefore \qquad\qquad \delta P \simeq \tfrac{19}{100}P = 19\% \text{ of } P$

Hence P will increase by 19% approximately.

Exercise 9.3

1) Show that an error of 1% in the measurement of the radius of a sphere leads to an error of approximately 2% in the outer surface area, and 3% in the volume.

2) Find the approximate percentage error in the calculated volume of a right circular cone if the radius is taken as 2% too large, and the height is taken as 3% too small.

3) A formula used in connection with helical springs is $y = \dfrac{8WnD^3}{Gd^4}$.

Find the percentage error in y if D is 1% too small, and d is 2% too large.

4) In the standard gas equation $\dfrac{pV}{T} = k$, the volume V is increased by 2%, the temperature T is diminished by 1%, and the value of constant k remains unaltered. What is the corresponding percentage change in the pressure p?

5) The deflection y at the centre of a steel rod of length l and circular cross-section of diameter d, simply supported at its ends and carrying a concentrated load W at its centre is given by the formula $y = \dfrac{Wl^3}{d^4}$. Find the percentage change in y when l increases by 2% and d decreases by 3%.

6) If x is so small that x^3 and all other higher powers of x can be neglected, find an approximation of $3\sqrt{(2 + x)}$.

7) If the height h of an isosceles triangle is small compared with the base of length $2b$, show that each of the slant sides has a length of approximately $b + h^2/2b$.

8) The volume per hour of water, V, passing through an injector of diameter d due to a pressure p is given by the formula $V = 2d^2p^{1/8}$. Find the approximate change in d if p is increased by 5% and V is decreased by 3.5%.

9) The resonant frequency of an oscillation in electrical circuits is given by the formula $f = 1/(2\pi\sqrt{LC})$. If the error in measuring L is 2%, and that in measuring C is 1%, calculate the maximum percentage error in calculating f.

10) A taut wire is held horizontally at two fixed points $2L$ inches apart. If the centre of the wire is deflected vertically by a small distance x, use the binomial theorem to show that the wire extends by an amount approximately equal to $x^2/2L$.

11) The period t of oscillation of a simple pendulum is given by the formula $t = 2\pi\sqrt{l/g}$ where l is the length and g the acceleration due to gravity. If, in an experiment, l is measured 1% too large and g is 5% too large, find the approximate error in t.

THE EXPONENTIAL FUNCTION

DEFINITION

The term *the exponential function* is generally used to mean the function e^x, where e is the base of Napierian logarithms.

THE EXPONENTIAL SERIES

The function e^x may be expanded as a power series and this is called the *exponential series*.

The exponential series is

$$e^x = 1 + x + \frac{x^2}{2!} + \frac{x^3}{3!} + \frac{x^4}{4!} + \frac{x^5}{5!} + \frac{x^6}{6!} + \ldots$$

CONVERGENCY

Suppose we wish to evaluate e^2. We put $x = 2$ into the series for e^x.

Thus

$$e^2 = 1 + 2 + \frac{2^2}{2!} + \frac{2^3}{3!} + \frac{2^4}{4!} + \frac{2^5}{5!} + \frac{2^6}{6!} + \ldots$$

106

$$= 1 + 2 + \frac{2 \times 2}{2 \times 1} + \frac{2 \times 2 \times 2}{3 \times 2 \times 1} + \frac{2 \times 2 \times 2 \times 2}{4 \times 3 \times 2 \times 1} + \frac{2 \times 2 \times 2 \times 2 \times 2}{5 \times 4 \times 3 \times 2 \times 1}$$

$$+ \frac{2 \times 2 \times 2 \times 2 \times 2 \times 2}{6 \times 5 \times 4 \times 3 \times 2 \times 1} + \ldots$$

$$= 1 + 2 + 2 + 1.3333 + 0.6667 + 0.2667 + 0.0889 + \ldots$$

Although initially the terms get larger, the fourth and subsequent terms become successively smaller. Thus the series is *convergent* and may be used to evaluate e^2 to any desired accuracy.

In general the exponential series is convergent for all values of x.

This means the series may be used to find the values of expressions such as $e^{3.1}$, e^{-4}, $e^{0.2}$, $e^{-1.56}$, etc. You may verify this by working through Question 1 in Exercise 10.1 at the end of this section.

EXAMPLE 10.1

Find a series for e^{-x}.

If we substitute $-x$ for x in the series for e^x, then

$$e^{-x} = 1 + (-x) + \frac{(-x)^2}{2!} + \frac{(-x)^3}{4!} + \frac{(-x)^4}{4!} + \frac{(-x)^5}{5!} + \frac{(-x)^6}{6!} + \ldots$$

or

$$e^{-x} = 1 - x + \frac{x^2}{2!} - \frac{x^3}{3!} + \frac{x^4}{4!} - \frac{x^5}{5!} + \frac{x^6}{6!} - \ldots$$

EXAMPLE 10.2

Find a series for $2e^3$.

If we substitute $x = 3$ in the series for e^x, then

$$e^3 = 1 + 3 + \frac{3^2}{2!} + \frac{3^3}{3!} + \frac{3^4}{4!} + \frac{3^5}{5!} + \ldots$$

$$\therefore \qquad 2e^3 = 2 \left[1 + 3 + \frac{3^2}{2!} + \frac{3^3}{3!} + \frac{3^4}{4!} + \frac{3^5}{5!} + \ldots \right]$$

The answer is better left in this form unless a numerical value is required.

EXAMPLE 10.3.

Find a series for $\frac{1}{2}(e^{4x} + e^{-4x})$.

If we substitute $4x$ for x in the series for e^x, then

$$e^{4x} = 1 + 4x + \frac{(4x)^2}{2!} + \frac{(4x)^3}{3!} + \frac{(4x)^4}{4!} + \frac{(4x)^5}{5!} + \ldots$$

and if we also substitute $-4x$ for x in the series for e^x, then:

$$e^{-4x} = 1 + (-4x) + \frac{(-4x)^2}{2!} + \frac{(-4x)^3}{3!} + \frac{(-4x)^4}{4!}$$

$$+ \frac{(-4x)^5}{5!} + \ldots$$

$$= 1 - 4x + \frac{(4x)^2}{2!} - \frac{(4x)^3}{3!} + \frac{(4x)^4}{4!} - \frac{(4x)^5}{5!} + \ldots$$

Now adding the series we have

$$e^{4x} + e^{-4x} = 1 + 4x + \frac{(4x)^2}{2!} + \frac{(4x)^3}{3!} + \frac{(4x)^4}{4!} + \frac{(4x)^5}{5!} + \ldots$$

$$+ 1 - 4x + \frac{(4x)^2}{2!} - \frac{(4x)^3}{3!} + \frac{(4x)^4}{4!} - \frac{(4x)^5}{5!}$$

$$+ \ldots$$

$$= 2[1] + 2\left[\frac{(4x)^2}{2!}\right] + 2\left[\frac{(4x)^4}{4!}\right] + 2\left[\frac{(4x)^6}{6!}\right] + \ldots$$

And dividing both sides by 2, we have

$$\frac{1}{2}(e^{4x} + e^{-4x}) = 1 + \frac{(4x)^2}{2!} + \frac{(4x)^4}{4!} + \frac{(4x)^6}{6!} + \ldots$$

THE NUMERICAL VALUE OF e

Although we have defined the exponential function as e^x, we still do not know how to arrive at the numerical value of the constant e.

This value may be found by putting $x = 1$ in the series for e^x.

We must also decide to what accuracy we require the numerical answer. This will determine how many terms we will have to include in our calculations. Suppose we decide on an answer correct to four decimal places. In the final rounding off we must make sure that the fifth decimal place is correct and this in turn will be affected by the sixth decimal place figures.

Thus

$$e = e^1 = 1 + 1 + \frac{1^2}{2!} + \frac{1^3}{3!} + \frac{1^4}{4!} + \frac{1^5}{5!} + \frac{1^6}{6!} + \frac{1^7}{7!} + \frac{1^8}{8!} + \frac{1^9}{9!} + \ldots$$

$$
\begin{aligned}
= \quad & 1.000\,000 \\
+ & 1.000\,000 \\
+ & 0.500\,000 \quad (\tfrac{1}{2} \text{ of the previous term}) \\
+ & 0.166\,667 \quad (\tfrac{1}{3} \text{ of the previous term}) \\
+ & 0.041\,667 \quad (\tfrac{1}{4} \text{ of the previous term}) \\
+ & 0.008\,333 \quad (\tfrac{1}{5} \text{ of the previous term}) \\
+ & 0.001\,389 \quad (\tfrac{1}{6} \text{ of the previous term}) \\
+ & 0.000\,198 \quad (\tfrac{1}{7} \text{ of the previous term}) \\
+ & 0.000\,025 \quad (\tfrac{1}{8} \text{ of the previous term}) \\
+ & 0.000\,003 \quad (\tfrac{1}{9} \text{ of the previous term})
\end{aligned}
$$

$$\therefore \qquad e = \quad 2.718\,282$$

Thus the value of e is 2.7183 correct to 4 decimal places.

THE UNIQUE PROPERTY $\frac{d}{dx}(e^x) = e^x$

We have

$$e^x = 1 + x + \frac{x^2}{2!} + \frac{x^3}{3!} + \frac{x^4}{4!} + \ldots$$

Now, if we differentiate this series term by term we obtain

$$\frac{d}{dx}(e^x) = 1 + \frac{2x}{2!} + \frac{3x^2}{3!} + \frac{4x^3}{4!} + \ldots$$

and considering a typical coefficient, for example,

$$\frac{4}{4!} = \frac{4}{4 \times 3 \times 2 \times 1} = \frac{1}{3!}$$

Thus

$$\frac{d}{dx}(e^x) = 1 + x + \frac{x^2}{2!} + \frac{x^3}{3!} + \ldots$$

which is the same as the original series.

Hence $\dfrac{d}{dx}(e^x) = e^x$ which confirms the result already obtained by a different method on p. 13.

This is the only mathematical function which does *not* change on differentiation.

Put another way we may say that for the graph of $y = e^x$ the rate of change $\dfrac{dy}{dx}$, at any point, is equal to e^x, i.e. $\dfrac{dy}{dx} = y$.

Exercise 10.1

The following examples should be solved using the exponential series. Numerical answers may be checked using a calculator.

1) Find, correct to three significant figures, the values of

(a) $e^{1.5}$ (b) $e^{3.1}$ (c) e^{-2} (d) $e^{-4.3}$

(e) $e^{0.2}$ (f) $e^{0.53}$ (g) $e^{-0.7}$

2) Find the first four terms of the series for

(a) e^{3x} (b) $e^{0.5x}$ (c) $e^{-1.3x}$ (d) $e^{-0.3x}$

3) Evaluate to three decimal places

(a) $3\left(e - \dfrac{1}{e}\right)$ (b) $\tfrac{1}{2}(e^{0.2} + e^{-0.2})$

4) Find the first four terms of the series for $\dfrac{e^{2x} + 1}{2e^x}$.

EXPONENTIAL GRAPHS

Curves of exponential functions which have equations of the type $y = e^x$ or $y = e^{-x}$ are called *exponential graphs*. We may plot the curves using values obtained from a calculator. For example, when $x = -2$, to find the value of e^{-2} the following sequence may be used:

$$\boxed{\text{AC}} \quad \boxed{2} \quad \boxed{+/-} \quad \boxed{e^x}$$

giving an answer 0.135 correct to three significant figures which is more than adequate for the purpose of plotting the graph. The two curves are shown in Fig. 10.1.

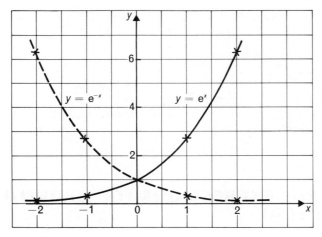

Fig. 10.1

In technology reference to the exponential curves is usually based on time. Thus the curves are plotted on a time (t) base and not an x base.

Now the graph of $y = e^t$ is called a *growth curve* and the graph of $y = e^{-t}$ is called a *decay curve*.

Practical applications based on growth and decay start from a given instant. This means we are dealing with actual (or real) times which, in numerical terms, are positive values of t. Thus only portions of the curves to the right of the vertical axis are generally used, as shown in Fig. 10.2.

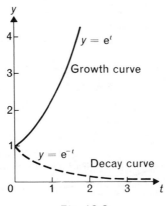

Fig. 10.2

Another form of growth has an equation of the form $y = 1 - e^{-t}$ and this curve is shown in Fig. 10.3.

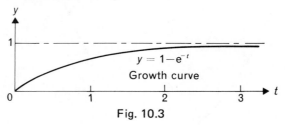

Fig. 10.3

GENERAL EXPRESSIONS FOR GROWTH AND DECAY

When used to solve problems the equations are modified to the following general forms

$$y = ae^{bt}, \quad y = ae^{-bt}, \quad \text{and} \quad y = a(1 - e^{-bt})$$

where a and b are constants which affect the initial value of y and also the rate of growth or decay. Sketches of the curves and some typical applications of each are shown in Figs. 10.4, 10.5 and 10.6.

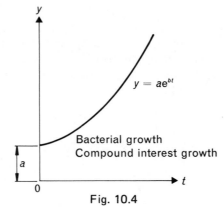

Fig. 10.4

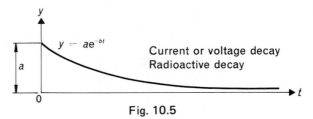

Fig. 10.5

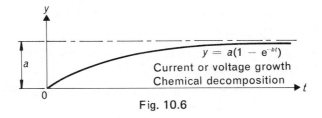

Fig. 10.6

COMPOUND INTEREST

Suppose we invest a capital sum of £40 for three years at compound interest of 60% per annum.

At the end of the first year the capital will have increased (in £ units) to

$$\text{Capital} + \text{Interest} = 40 + \tfrac{60}{100} \times 40 = 40(1 + \tfrac{60}{100})$$

Thus the capital has increased by a factor of $(1 + \tfrac{60}{100})$ during one year.

Now our capital sum at the beginning of the second year will be $40(1 + \tfrac{60}{100})$ and this in turn will increase by a factor of $(1 + \tfrac{60}{100})$ during the year.

At the end of the second year the capital will have increased to

$$40(1 + \tfrac{60}{100}) \times (1 + \tfrac{60}{100}) \quad \text{or} \quad 40(1 + \tfrac{60}{100})^2$$

Similarly at the end of the third year our capital will have increased to

$$40(1 + \tfrac{60}{100})^3 \quad \text{which is} \quad £163.84$$

As an alternative to the above suppose that the interest is paid half-yearly. This is equivalent to 30% with a total of 6 interest payment intervals. Using the above ideas our final sum will be

$$40(1 + \tfrac{30}{100})^6 \quad \text{or} \quad £193.07$$

Another alternative would be quarterly interest payments which would be equivalent to 15% with a total of 12 interest payment intervals. Thus our final sum will be

$$40(1 + \tfrac{15}{100})^{12} \quad \text{or} \quad £214.01$$

We can see that as the length of the interest payment intervals are reduced so our final sum is increased. Ideally we would like the length of the interest payment intervals to be very small indeed.

This would approach 'continuous' payments which would result in obtaining the 'best possible accumulated sum'.

In this case after n years our best possible accumulated sum may be found from the expression $40e^{n(60/100)}$.

Thus at the end of 3 years the best possible accumulated sum

$$= 40e^{3(60/100)} = 40e^{1.8} \quad \text{or} \quad £241.99$$

The results of the above calculations are displayed in Fig. 10.7.

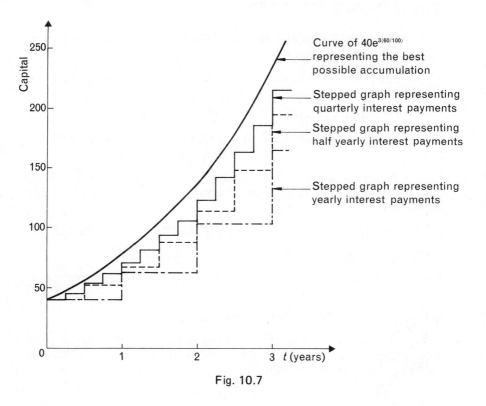

Fig. 10.7

Capital growths resulting from quarterly, half-yearly and yearly compound interest payment intervals are shown by the stepped graphs. The curve is that of the expression $40e^{3(60/100)}$ representing the best possible accumulated sum. You will see that as the interest payment intervals reduce so the top left-hand peaks of each step move closer to the curve. The smaller the interest interval payment, the closer to the curve the result will be. You may like to try this for $\frac{1}{10}$ yearly intervals at 6% interest.

In practice the best possible accumulated sum is never reached. However, interest paid on deposit accounts at banks, usually calculated on a daily basis, gives capital build-up closely approaching the exponential curve.

In general, if a capital sum (or principal) P is invested for n years at compound interest $r\%$ per annum, then the best possible accumulated sum is given by the expression

$$Pe^{nr/100}$$

The corresponding capital after n years for yearly interest interval payments is given by

$$P\left(1 + \frac{r}{100}\right)^n$$

OTHER EXAMPLES IN SCIENCE AND TECHNOLOGY

EXAMPLE 10.4

The population size, N, at a certain time, t hours, after commencement of growth of a unit-sized population, is given by the exponential growth relationship $N = e^{0.8t}$. Show that the instantaneous rate of growth is proportional to the population size.

The values of N for corresponding values of t may be found using the scientific calculator. The table of values shows results for values of t from zero to 4 hours.

t hours	0	0.5	1	1.5	2	2.5	3	3.5	4
$N = e^{0.8t}$	1	1.49	2.23	3.32	4.95	7.39	11.0	16.4	24.5

The curve is shown plotted in Fig. 10.8.

The *instantaneous rate of growth* means *the rate of growth at any instant*. This is given by the gradient of the curve at any particular point. The gradient may be found by drawing a tangent to the curve at the point and calculating its slope by constructing a suitable right-angled triangle.

At P, using the right-angled triangle shown:

$$\text{the gradient} = \frac{5.4}{3} = 1.8 \quad \text{and the value of } N = 2.23$$

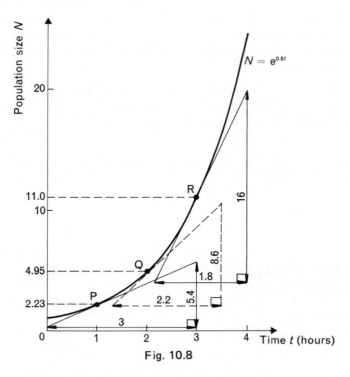

Fig. 10.8

Hence   the ratio $\dfrac{\text{gradient}}{N \text{ value}} = \dfrac{1.8}{2.23} = 0.81$

Similarly at Q

the ratio $\dfrac{\text{gradient}}{N \text{ value}} = \dfrac{8.6/2.2}{4.95} = 0.79$

and at R   the ratio $\dfrac{\text{gradient}}{N \text{ value}} = \dfrac{16/1.8}{11} = 0.81$

You may like to plot the curve and calculate the ratio $\dfrac{\text{gradient}}{N \text{ value}}$ for other points. We can see from these results that the value of the ratio (within the limitations of accuracy of values obtained from the graph) is constant — in this case 0.80

It is, therefore, reasonable to assume that at any point on the exponential curve the ratio

$$\dfrac{\text{gradient}}{N \text{ value}} = \text{constant}$$

or rearranging   gradient $=$ (constant)(N value)

∴   gradient $\propto$ N value

Alternatively using differentiation:

If $N = e^{0.8t}$

then $\dfrac{dN}{dt} = 0.8e^{0.8t}$

or $\dfrac{dN}{dt} = 0.8N$

Thus $\dfrac{dN}{dt} \propto N$

Thus the instantaneous rate of growth is proportional to the population size.

> From this we may conclude that a property of an exponential curve is that, at any point, the gradient is proportional to the N value (i.e. the ordinate).

EXAMPLE 10.5

The amount of radioactivity remaining, N, is given by the equation $N = N_0 e^{-0.231t}$, where N_0 is the initial intensity of radioactivity and t is the time in years. Draw the decay curve and use it to find:

a) the half-life (the time for half the radioactivity to decay),

b) how much radioactivity there is remaining after 6 years.

Fig. 10.9 shows the curve of N against t, for values of t from 0 to 8 seconds, and values of N which are decimal fractions of N_0.

If for the first value we choose, say $t = 8$, to find the value of $e^{-0.231 \times 8}$ we use the calculator sequence

| AC | 0 | . | 2 | 3 | 1 | +/− | Min |

| × | 8 | = | e^x |

We may benefit from putting -0.231 in the memory. Subsequent values of $e^{-0.231t}$ may then be found using the sequence

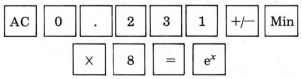

| t value | × | MR | = | e^x |

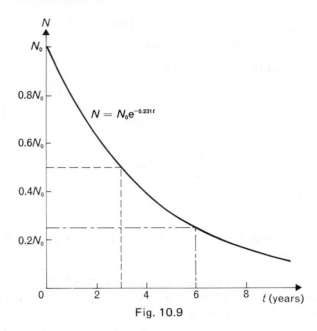

Fig. 10.9

a) At half-life there is half the amount of radioactivity remaining. This is when $N = 0.5N_0$ and from the graph this occurs when $t = 3.0$ years.

b) Also from the graph when $t = 6$ years we may read off the corresponding value of N which is $0.25N_0$, which means that one quarter of the radioactivity remains.

EXAMPLE 10.6

The formula $i = 2(1 - e^{-10t})$ gives the relationship between the instantaneous current i amperes and the time t seconds in an inductive circuit. Plot a graph of i against t taking values of t from 0 to 0.3 seconds at intervals of 0.05 seconds. Hence find:

a) the initial rate of growth of the current i when $t = 0$, and

b) the time taken for the current to increase from 1 to 1.6 amperes.

The curve is shown plotted in Fig. 10.10 from values obtained from the sequence of operations,

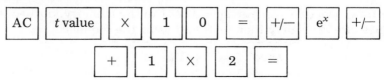

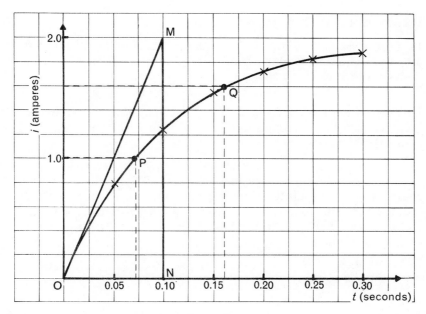

Fig. 10.10

a) When $t = 0$ the initial rate of growth will be given by the gradient of the tangent at O. The tangent at O is the line OM and its gradient may be found by using a suitable right-angled triangle at MNO and finding the ratio $\dfrac{MN}{ON}$.

Hence the initial rate of growth of $i = \dfrac{MN}{ON} = \dfrac{2 \text{ amperes}}{0.1 \text{ seconds}}$

$= 20 \text{ amperes per second.}$

b) The point P on the curve corresponds to a current of 1.0 amperes and the time at which this occurs may be read from the t scale and is 0.07 seconds.

Similarly point Q corresponds to a 1.6 ampere current and occurs at 0.16 seconds.

Hence the time between P and Q is $0.16 - 0.07 = 0.09$ seconds.

This means that the time for the current to increase from 1 to 1.6 amperes is 0.09 seconds.

Exercise 10.2

1) Plot a graph of $y = 3e^{2t}$ for values of t from 0 to 2 at 0.25 unit intervals. Use the graph to find the value of y when $t = 0.3$, and the value of t when $y = 5.4$.

2) Using values of x from 0 to $+4$ at half unit intervals plot a graph of $y = 2e^{-3t}$. Hence find the value of t when $y = 2$, and the gradient of the curve when $t = 2$.

3) For a constant pressure process on a certain gas the formula connecting the absolute temperature T and the specific entropy s is $T = 24e^{3s}$. Plot a graph of T against s taking values of s equal to 1.000, 1.033, 1.066, 1.100, 1.133, 1.166 and 1.200. Use the graph to find the value of:

(a) T when $s = 1.09$;

(b) s when $T = 700$.

4) The equation $i = 2.4e^{-6t}$ gives the relationship between the instantaneous current, i mA, and the time, t seconds.

Plot a graph of i against t for values of t from 0 to 0.6 seconds at 0.1 second intervals.

Use the curve obtained to find the rate at which the current is decreasing when $t = 0.2$ seconds.

5) In a capacitive circuit the voltage v and the time t seconds are connected by the relationship $v = 240(1 - e^{-5t})$.

Draw the curve of v against t for values of $t = 0$ to $t = 0.7$ seconds at 0.1 second intervals.

Hence find:

(a) the time when the voltage is 140 volts; and

(b) the initial rate of growth of the voltage when $t = 0$.

6) The number of cells, N, in a bacterial population in time, t hours, from the commencement of growth is given by $N = 100e^{1.7t}$. Find:

(a) the size of the population after 4 hours growth;

(b) the time in which the population increases ten-fold from its initial value;

(c) the instantaneous rate of growth after 3 hours from the start.

7) Given that the mass, m grams, of a bacterial population after t hours from the beginning of growth is given by $m = (10^{-10})e^{1.2t}$, find:

(a) the mass of the population after 2 hours from growth commencement;

(b) the mass of the population at beginning of growth;

(c) the time when the population has doubled from its initial value.

8) The decomposition of a chemical compound, C, over a period of time, t, is given by $C = k(1 - e^{-0.2t})$. If $k = 10$ find the rate of decomposition after 15 seconds.

LOGARITHMIC SCALES

If has been shown earlier that:

$$\text{Number} = (\text{base})^{\text{logarithm}}$$

and if we use a base of 10 then:

$$\text{Number} = (10)^{\text{logarithm}}$$

Since

$\qquad 1000 = 10^3 \qquad$ then we may write $\quad \log_{10}1000 = 3$

and since

$\qquad 100 = 10^2 \qquad$ then we may write $\quad \log_{10}100 = 2$

and since

$\qquad 10 = 10^1 \qquad$ then we may write $\quad \log_{10}10 = 1$

and since

$\qquad 1 = 10^0 \qquad$ then we may write $\quad \log_{10}1 = 0$

and since

$\qquad 0.1 = 10^{-1} \quad$ then we may write $\quad \log_{10}0.1 = -1$

and since

$\qquad 0.01 = 10^{-2} \quad$ then we may write $\quad \log_{10}0.01 = -2$

and since

$\qquad 0.001 = 10^{-3} \quad$ then we may write $\quad \log_{10}0.001 = -3$

and so on.

These logarithms may be shown on a scale as shown in Fig. 10.11.

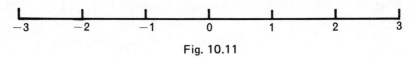

Fig. 10.11

However, since we wish to plot numbers directly on to the scale (without any reference to their logarithms), the scale is labelled as shown in Fig. 10.12.

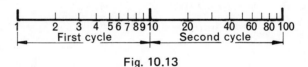

Fig. 10.12

Each division is called a cycle and is sub-divided using a logarithmic scale as, for instance, the scales on a slide rule. Two such cycles are shown in Fig. 10.13.

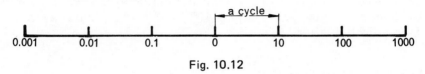

Fig. 10.13

The choice of numbers on the scale depends on the numbers allocated to the variable in the problem to be solved. Thus in Fig. 10.13 the numbers run from 1 to 100.

Consider the following relationships in which z and t are the variables, whilst a, b, and n are constants.

Also e is the 'exponential e' which is, of course, a constant.

$z = ab^t$

Now $$z = ab^t$$

and taking logs $$\log z = \log(a \times b^t)$$
$$= \log b^t + \log a$$

∴ $$\log z = (\log b)t + \log a$$

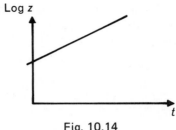

Fig. 10.14

The given values of the variables will satisfy this equation if they satisfy the original equation. Comparing this equation with $y = mx + c$, which is the standard equation of a straight line, we see that if we plot $\log z$ on the y-axis and t on the x-axis the result will be a straight line (Fig. 10.14) and the values of the constants n and a may be found using the two-point method.

$z = ae^{bt}$

Now
$$z = ae^{bt}$$

Taking logs gives
$$\log z = \log(a \times e^{bt})$$
$$= \log e^{bt} + \log a$$

$\therefore$
$$\log z = (b \log e)t + \log a$$

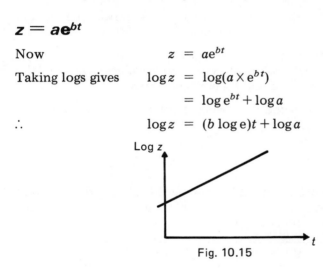

Fig. 10.15

Again proceeding in a manner similar to that used for the previous equation, we plot $\log z$ on the y-axis and t on the x-axis and again obtain a straight line (Fig. 10.15).

SEMI-LOGARITHMIC GRAPH PAPER

Logarithmic scales may be used on graph paper in place of the more usual linear scales. By using graph paper ruled in this way log plots may be made without the necessity of finding the logs of each given value.

Semi-logarithmic (or log–linear) graph paper shown in Fig. 10.16 has logarithmic scales on the vertical axis, but the usual linear scale on the horizontal axis.

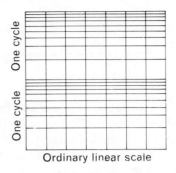

Fig. 10.16

The use of semi-logarithmic graph paper enables power or exponential relationships between two variables to be verified quickly.

A straight line graph on semi-logarithmic (or log–linear) graph paper indicates a relationship between the variables z and t of the form $z = ab^t$ or $z = ae^{bt}$, a and b being constants.

The use of these scales and the special graph paper is shown by the examples which follow.

EXAMPLE 10.7

The table gives values obtained in an experiment. It is thought that the law may be of the form $z = ab^t$, where a and b are constants. Verify this and find the law.

t	0.190	0.250	0.300	0.400
z	11 220	18 620	26 920	61 660

We think that the relationship is of the form

$$z = ab^t$$

which gives (see text)

$$\log z = (\log b)t + \log a \qquad [1]$$

Hence we plot the given values of z on a vertical log scale — the t values however will be on the horizontal axis on an ordinary linear scale (Fig. 10.17).

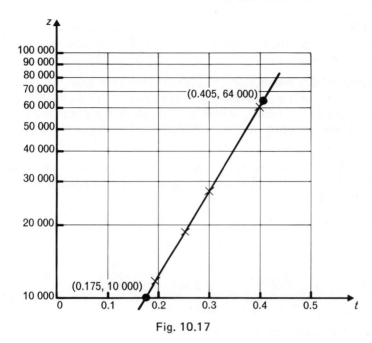

Fig. 10.17

The points lie on a straight line, and therefore the given values of z and t obey the law.

We now have to select the coordinates of two points lying on the line.

Point (0.405, 64 000) lies on the line.

Substituting in equation [1] gives

$$\log 64\,000 \; = \; (\log b)0.405 + \log a \qquad [2]$$

Point (0.175, 10 000) lies on the line.

Substituting in equation [1] gives

$$\log 10\,000 \; = \; (\log b)0.175 + \log a \qquad [3]$$

Subtracting equation [3] from equation [2] gives

$$\log 64\,000 - \log 10\,000 \; = \; (\log b)(0.405 - 0.175)$$

$\therefore$
$$\log \left(\frac{64\,000}{10\,000}\right) \; = \; 0.230 \,(\log b)$$

or
$$\log 6.4 \; = \; 0.230 \log b$$

At this stage of our solution we have to decide whether to proceed using common logarithms (to the base 10) or natural logarithms (to the base e).

At first sight it seems that common logarithms must be used, since these were used in establishing the layout of the logarithmic scales on the graph. However, the actual values of the coordinates of the two chosen points were independent of the logarithmic base chosen.

The solution here uses common logarithms. You may find it instructive to work through using natural logarithms and verify that the same results are obtained.

Thus we have

$$\log_{10} 6.4 = 0.230 \log_{10} b$$

or
$$\log_{10} b = \frac{\log_{10} 6.4}{0.230} = \frac{0.8062}{0.230} = 3.505$$

$$\therefore \qquad b = 3200$$

Now substituting in equation [3] gives

$$\log_{10} 10\,000 = (3.505)0.175 + \log_{10} a$$

$$\therefore \qquad 4 = 0.613 + \log_{10} a$$

$$\therefore \qquad \log_{10} a = 4 - 0.613 = 3.387$$

$$\therefore \qquad a = 2440$$

Hence the law is $z = 2440(3200)^t$

EXAMPLE 10.8

V and t are connected by the law $V = ae^{bt}$. If the values given in the table satisfy the law, find the constants a and b.

t	0.05	0.95	2.05	2.95
V	20.70	24.49	30.27	36.06

The law is $V = ae^{bt}$

which gives (see text)

$$\log V = (b \log e)t + \log a \qquad [1]$$

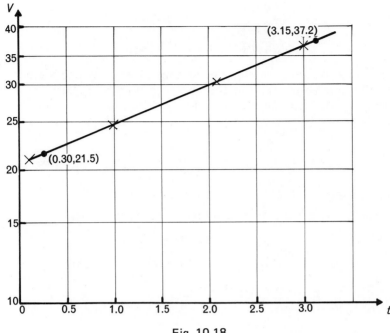

Fig. 10.18

As in the last example V values are plotted on a log scale on the vertical axis, while the t values are plotted on the horizontal axis on an ordinary linear scale (Fig. 10.18).

Since the points lie on a straight line, the given values of t and V satisfy the given law.

We now have to select two points *which lie* on the line and determine their coordinates.

Point (3.15, 37.2) lines on the line.

Substituting in equation [1] gives

$$\log 37.2 = (b \log e)(3.15) + \log a \qquad [2]$$

Point (0.30, 21.5) lies on the line.

Substituting in equation [1] gives

$$\log 21.5 = (b \log e)(0.30) + \log a \qquad [3]$$

Subtracting gives

$$\log 37.2 - \log 21.5 = (b \log e)(3.15 - 0.30)$$

$$\therefore \qquad \log \frac{37.2}{21.5} = (b \log e)(2.85)$$

or $\qquad \log 1.73 = 2.85 \times b \times \log e$

As in the previous example, we must now decide which log base to use. Where $\log e$ is involved the choice of natural logarithms helps to simplify the calculations since $\log_e e = 1$.

Thus $\qquad \log_e 1.73 = 2.85 \times b \times 1$

$$\therefore \qquad b = \frac{\log_e 1.73}{2.85} = \frac{0.548}{2.85} = 0.192$$

Now putting $b = 0.192$ into equation [3] gives

$$\log_e 21.5 = 0.192 \times 1 \times 0.30 + \log_e a$$

$\therefore \qquad 3.068 = 0.0576 + \log_e a$

$\therefore \qquad \log_e a = 3.068 - 0.0576 = 3.01$

from which $\qquad a = 20.3$

Exercise 10.3

1) The values given in the following table are thought to obey a law of the type $y = ab^{-x}$. Check this statement and find the values of the constants a and b.

x	0.1	0.2	0.4	0.6	1.0	1.5	2.0
y	175	122	60	32	6.4	1.28	0.213

2) The force F on the tight side of a driving belt was measured for different values of the angle of lap θ and the following results were obtained:

F	7.4	11.0	17.5	24.0	36.0
θ rad	$\pi/4$	$\pi/2$	$3\pi/4$	π	$5\pi/4$

Construct a graph to show these values conform approximately to an equation of the form $F = ke^{\mu\theta}$. Hence find the constants μ and k.

3) A particular bacterial growth, checked at regular time intervals, is shown in the table.

Bacteria ($\times 10^3$)	2.90	3.37	3.92	4.56	5.03	5.56	6.80
Time, t (minute)	15	30	45	60	70	80	100

The bacteria, B, and the time, t, are thought to be related by $B = Ce^{kt}$ where C and k are constants. With the aid of a suitable graph check that this is correct and evaluate the constants.

4) Height, h, and atmospheric pressure, p, are thought to be related by the law $pe^{ah} = c$ where a and c are constants. Using the table below test the correctness of this statement and if it is true evaluate a and c.

h (m)	0	2000	4000	6000	8000	10 000
p (cm of mercury)	76.2	66.7	58.4	51.1	44.7	39.1

5) The table below shows the variation in the coefficient of viscosity of a particular fluid, z, at various temperatures, t.

t (°C)	0	6	12	18	24
z	40.0	23.3	13.6	7.9	4.6

By plotting a suitable straight line graph verify that z and t are related by the equation $z = Ae^{-at}$ and hence find the values of the constants a and A.

6) Across a condensor, after the supply has been cut, the voltage, V, and the time, t, are thought to be related by the law $V = V_0 e^{-at}$ as in the table below.

Time, t (second)	5	10	15	20	25	30
Voltage, V (volt)	4.00	3.18	2.55	2.03	1.62	1.30

By plotting the appropriate graph show this to be true and hence evaluate the constants, V_0 and a.

7) A capacitor and resistor are connected in series. The current i amperes after time t seconds is thought to be given by the equation $i = Ie^{-t/T}$ where I amperes is the initial charging current and T seconds is the time constant. Using the following values verify the relationship and find the values of the constants I and T.

i amperes	0.0156	0.0121	0.009 45	0.007 36	0.005 73
t seconds	0.05	0.10	0.15	0.20	0.25

11. GRAPHS OF NON-LINEAR RELATIONSHIPS

After reaching the end of this chapter you should be able to:

1. *Draw up suitable tables of values and plot the curves of the types:*

   $y = ax^2 + bx + c,$ $y = \dfrac{a}{x},$ $y = x^{1/2},$

   $x^2 + y^2 = a^2,$ $\dfrac{x^2}{a^2} + \dfrac{y^2}{b^2} = 1,$ $\dfrac{x^2}{a^2} - \dfrac{y^2}{b^2} = 1,$

   $xy = c^2.$

2. *Recognise the effects on the above curves by changes in the constants a, b, and c.*

3. *Reduce non-linear physical laws, such as*

   $y = ax^2 + b$ *or* $y = \dfrac{a}{x} + b,$ *and* $y = ax^n$

   (using logarithms), to a straight line graph form.

4. *Plot the corresponding straight line graph to verify the law, determine the values of the constants a and b, and find intermediate values.*

5. *Use full logarithmic graph paper with equations of the type z = ax^n.*

GRAPHS OF FAMILIAR EQUATIONS

It is important for technicians to be able to recognise the shapes and layouts of curves related to their equations. In this section we will draw graphs of the more common equations.

Graph of $y = ax^2 + bx + c$: Parabola

The important part of the curve is usually the portion in the vicinity of the vertex of the parabola (Fig. 11.1).

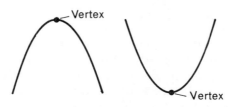

Fig. 11.1

The shape and layout will depend on the values of the constants *a*, *b* and *c* and we will examine the effect of each constant in turn.

130

Constant *a*

Consider the equation $y = ax^2$. The table of values given below is for $a = 4$, $a = 2$ and $a = -1$:

x	-3	-2	-1	0	1	2	3
$y = 4x^2$	36	16	4	0	4	16	36
$y = 2x^2$	18	8	2	0	2	8	18
$y = -x^2$	9	4	1	0	1	4	9

Fig. 11.2 shows the graphs of $y = ax^2$ when $a = 4$, $a = 2$ and $a = -1$

We can see that if the value of *a* is positive the curve is shaped ⌣, and the greater the value of *a*, the 'steeper' the curve rises.

Negative values of *a* give a curve shaped ⌢.

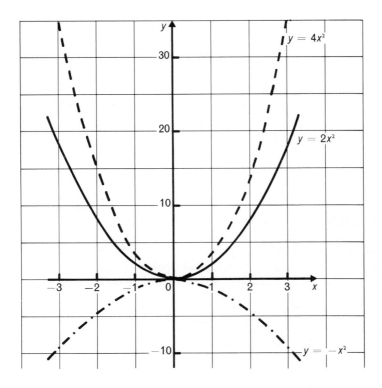

Fig. 11.2

Constant b

Consider the equation $y = x^2 + bx$. The table of values given below is for $b = 2$ and $b = -3$:

x	-3	-2	-1	0	1	2	3	4
$y = x^2 + 2x$	3	0	-1	0	3	8	15	24
$y = x^2 - 3x$	18	10	4	0	-2	-2	0	4

Fig. 11.3 shows the graphs of $y = x^2 + bx$ when $b = 2$ and $b = -3$. The effect of a positive value of b is to move the vertex to the left of the vertical y-axis, whilst a negative value of b moves the vertex to the right of the vertical axis.

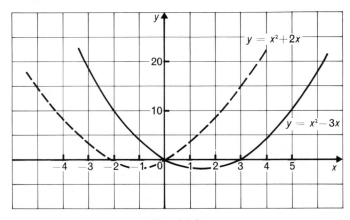

Fig. 11.3

Constant c

Consider the equation $y = x^2 - 2x + c$. The table of values given below is for $c = 10$, $c = 5$, $c = 0$ and $c = -5$

x	-3	-2	-1	0	1	2	3	4
$y = x^2 - 2x + 10$	25	18	13	10	9	10	13	18
$y = x^2 - 2x + 5$	20	13	8	5	4	5	8	13
$y = x^2 - 2x$	15	8	3	0	-1	0	3	8
$y = x^2 - 2x - 5$	10	3	-2	-5	-6	-5	-2	3

Fig. 11.4 shows the graphs of $y = x^2 - 2x + c$ when $c = 10$, $c = 5$, $c = 0$ and $c = -5$. As we can see the effect is to move the vertex up or down according to the magnitude of c.

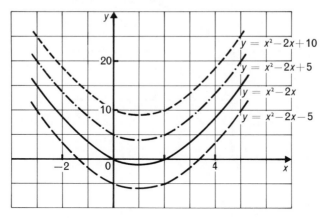

Fig. 11.4

The graph of $y = ax^{\frac{1}{2}}$ or $y = a\sqrt{x}$:
Parabola (with horizontal axis)

The table of values given below is for $a = 4$ which is the graph of $y = 4\sqrt{x}$:

x	All negative values	0	1	2	3	4	5	6
$y = 4\sqrt{x}$	No real values	0	± 4	± 5.66	± 6.93	± 8	± 8.94	± 9.80

The graph of $y = 4\sqrt{x}$ is shown in Fig. 11.5. We can see that as the value of a is increased so the 'overall depth' of the figure increases.

If the given equation is rearranged we may obtain

$$x = \frac{y^2}{a^2}$$

or

$$x = (\text{A constant}) \times y^2$$

which is the equation of a parabola with a horizontal axis of symmetry.

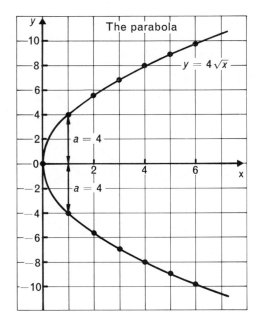

Fig. 11.5

The graph of $y = \frac{a}{x}$:
Reciprocal curve (rectangular hyperbola)

The table of values given below is for $a = 4$ which is the graph of $y = \frac{4}{x}$:

y	-10	-8	-6	-4	-2	-1	-0.5	0
$y = \dfrac{4}{x}$	-0.4	-0.5	-0.67	-1	-2	-4	-8	∞
	Similar numerical results will occur for positive values							

The graph of $y = \dfrac{4}{x}$ is shown in Fig. 11.6. We can see that as the value of a decreases the curves are brought nearer to the axes, and vice versa.

We should also note that for extreme values the curves will become nearer and nearer to the axes, but will never actually 'touch' them. At these extreme values the axes are said to be *asymptotic* to the curves. In this case the axes are also called *asymptotes*.

We shall see later on page 137 that the curves comprise, what is called, a *rectangular hyperbola*.

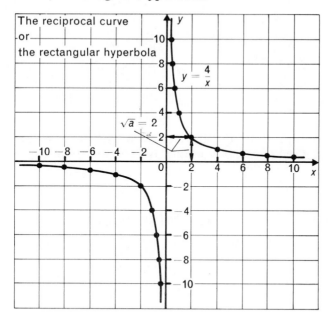

Fig. 11.6

The graph of $x^2 + y^2 = a^2$: Circle

The graph of $x^2 + y^2 = 4$, which is when $a = 2$, is shown in Fig. 11.7.

The table of values has been omitted here but you may find it useful to rearrange the equation and plot the curve.

We can see that the radius of the circle is given by the value of constant a.

The circle

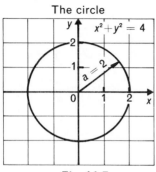

Fig. 11.7

The ellipse

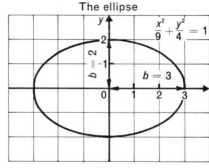

Fig. 11.8

The graph of $\dfrac{x^2}{a^2}+\dfrac{y^2}{b^2}=1$: Ellipse

The graph of $\dfrac{x^2}{9}+\dfrac{y^2}{4}=1$, which is when $a=3$ and $b=2$, is shown in Fig. 11.8 (see p. 135).

The line along which the greatest dimension (given by $2a$) across the ellipse is measured is called the *major axis*. In this case it lies along the x-axis.

The line along which the least dimension (given by $2b$) across the ellipse is measured is called the *minor axis*. In this case it lies along the y-axis.

The graph of $\dfrac{x^2}{a^2}-\dfrac{y^2}{b^2}=1$: Hyperbola

The graph of $\dfrac{x^2}{9}-\dfrac{y^2}{4}=1$, which is when $a=3$ and $b=2$, is shown in Fig. 11.9.

The ratio $\pm\dfrac{b}{a}$, in this case $\pm\dfrac{2}{3}$, gives the gradients of two straight lines called asymptotes. At extreme values these lines will become nearer and nearer to the curves but will never actually 'touch' them. They are thus *asymptotic* to the curves.

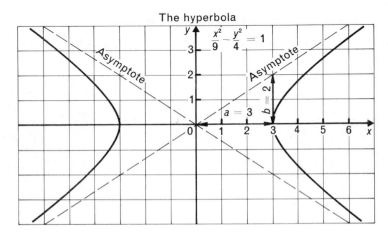

Fig. 11.9

The graph of $xy = c^2$:
Rectangular hyperbola

The graph of $xy = 9$, which is when $c = 3$, is shown in Fig. 11.10.

We can see that the curve is similar to the reciprocal curve discussed earlier. It is also similar to the ordinary hyperbola, the description *rectangular* referring to the fact that the asymptotes are also the rectangular axes.

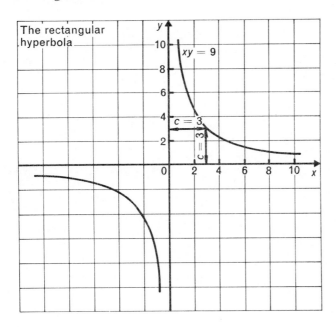

Fig. 11.10

Exercise 11.1

State which answer or answers are correct in Questions 1-9. In every diagram the origin is at the intersection of the axes.

1) The graph of $y = x^2$ is:

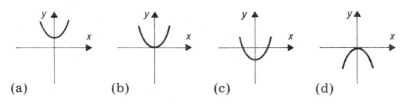

(a)   (b)   (c)   (d)

2) The graph of $y = x^2 + 2$ is:

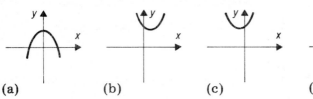 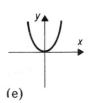

(a)   (b)   (c)   (d)   (e)

3) The graph of $y = -2x^2$ is:

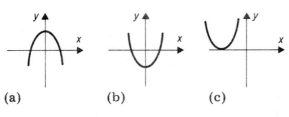

(a)   (b)   (c)   (d)   (e)

4) The graph of $y = 4 - x^2$ is:

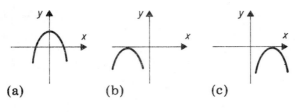

(a)   (b)   (c)   (d)   (e)

5) The graph of $y = x^2 - 3x + 2$ is:

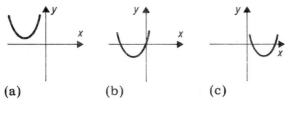

(a)   (b)   (c)   (d)   (e)

6) The graph of $y = x^2 + 4x + 4$ is:

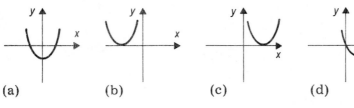

(a)   (b)   (c)   (d)   (e)

7) The graph of $\dfrac{x^2}{16} + \dfrac{y^2}{9} = 1$ is:

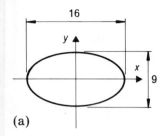

(a)

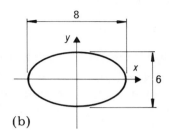

(b)

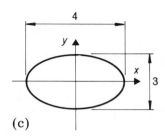

(c)

8) The graph of $\dfrac{x^2}{9} - \dfrac{y^2}{4} = 1$ is:

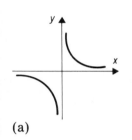

(a)

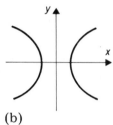

(b)

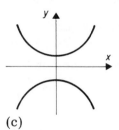

(c)

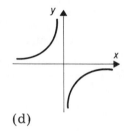

(d)

9) The graph of $xy = 9$ is:

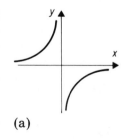

(a)

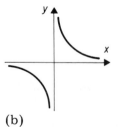

(b)

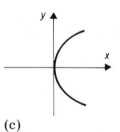

(c)

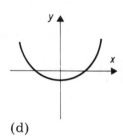

(d)

10) What is the equation of the circle shown in the diagram?

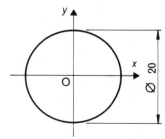

11) The profile of the cross-section of a headlamp reflector is parabolic in shape, having an equation $y = 0.00617x^2$. Plot the profile for values of x from 0 to 90 mm.

12) An elliptical template has an equation: $\dfrac{x^2}{3600} + \dfrac{y^2}{1600} = 1$

where x and y have mm units. Plot the shape of the template.

13) The pressure, $p\,\mathrm{MN/m^2}$, and volume, $V\,\mathrm{m^3}$, of a fixed mass of air at a constant temperature are connected by the equation: $pV = 0.16$. Plot the graph of pressure against volume, as the volume changes from 0 to $1\,\mathrm{m^3}$.

NON-LINEAR LAWS WHICH CAN BE REDUCED TO THE LINEAR FORM

Many non-linear equations can be reduced to the linear form by making a suitable substitution.

Common forms of non-linear equations are (a and b constants):

$y = \dfrac{a}{x} + b$	$y = \dfrac{a}{x^2} + b$	$y = ax^2 + b$	$y = a\sqrt{x} + b$

$y = \dfrac{a}{\sqrt{x}} + b$

Consider $y = \dfrac{a}{x} + b$

Let $z = \dfrac{1}{x}$ so that the equation becomes $y = az + b$. If we now plot values of y against the corresponding values of z we will get a straight line since $y = az + b$ is of the standard linear form. In effect y has been plotted against $\dfrac{1}{x}$. The following example illustrates this method.

EXAMPLE 11.1

An experiment connected with the flow of water over a rectangular weir gave the following results:

C	0.503	0.454	0.438	0.430	0.425	0.421
H	0.1	0.2	0.3	0.4	0.5	0.6

The relation between C and H is thought to be of the form $C = \dfrac{a}{H} + b$. Test if this is so and find the values of the constants a and b.

By putting $z = \dfrac{1}{H}$ then the equation $C = \dfrac{a}{H} + b$ becomes $C = az + b$ which is the standard form of a straight line. Hence by plotting C against $\dfrac{1}{H}$ we should get a straight line if the law is true. To try this we draw up a table showing corresponding values of C and $\dfrac{1}{H}$.

C	0.503	0.454	0.438	0.430	0.425	0.421
$z = \dfrac{1}{H}$	10.00	5.00	3.33	2.50	2.00	1.67

The graph obtained is shown in Fig. 11.11. It is a straight line and hence the given values follow a law of the form $C = \dfrac{a}{H} + b$.

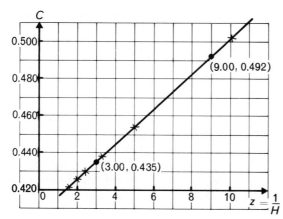

Fig. 11.11

To find the values of a and b we choose two points which lie on the straight line.

The point $(3.00, 0.435)$ lies on the line.

$\therefore$ $\qquad\qquad\qquad 0.435 = 3.00a + b$ $\qquad\qquad$ [1]

The point $(9.00, 0.492)$ also lies on the line.

$\therefore$ $\qquad\qquad\qquad 0.492 = 9.00a + b$ $\qquad\qquad$ [2]

Subtracting equation [1] from equation [2] gives

$$0.492 - 0.435 = a(9.00 - 3.00)$$

$$\therefore \qquad a = 0.0095$$

Substituting this value for a in equation [1] gives

$$0.435 = 3.00 \times 0.0095 + b$$

$$\therefore \qquad b = 0.435 - 0.0285 = 0.407$$

Hence the values of a and b are 0.0095 and 0.407 respectively.

Consider $y = ax^2 + b$

Let $z = x^2$ and, as previously, if we plot values of y against z (in effect x^2) we will get a straight line since $y = az + b$ is of the standard form. The following example illustrates this method.

EXAMPLE 11.2

The fusing current I amperes for wires of various diameters d mm is as shown below:

d (mm)	5	10	15	20	25
I (amperes)	6.25	10	16.25	25	36.25

It is suggested that the law $I = ad^2 + b$ is true for the range of values given, a and b being constants. By plotting a suitable graph show that this law holds and from the graph find the constants a and b. Using the values of these constants in the equation $I = ad^2 + b$ find the diameter of the wire required for a fusing current of 12 amperes.

By putting $z = d^2$ the equation $I = ad^2 + b$ becomes $I = az + b$ which is the standard form of a straight line. Hence by plotting I against d^2 we should get a straight line if the law is true. To try this we draw up a table showing corresponding values of I and d^2.

$z = d^2$	25	100	225	400	625
I	6.25	10	16.25	25	36.25

From the graph (Fig. 11.12) we see that the points do lie on a straight line and hence the values obey a law of the form $I = ad^2 + b$.

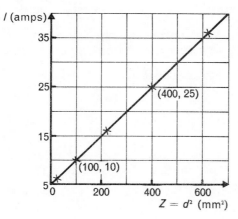

Fig. 11.12

To find the values of a and b choose two points which lie on the line and find their coordinates.

The point $(400, 25)$ lies on the line

$$\therefore \qquad 25 = 400a + b \qquad\qquad [1]$$

The point $(100, 10)$ lies on the line

$$\therefore \qquad 10 = 100a + b \qquad\qquad [2]$$

Subtracting equation [2] from equation [1] gives

$$15 = 300a$$

$$\therefore \qquad a = 0.05$$

Substituting $a = 0.05$ in equation [2] gives

$$10 = 100 \times 0.05 + b$$

$$\therefore \qquad b = 5$$

Therefore the law is

$$I = 0.05d^2 + 5$$

When $I = 12$

$$12 = 0.05d^2 + 5$$

$$\therefore \qquad d = \sqrt{140} = 11.8\,\text{mm}$$

Consider $\quad y = \dfrac{a}{x^2} + b$

Let $z = \dfrac{1}{x^2}$ so that the equation becomes $y = az + b$. If we now plot values of y against corresponding values of z we will get a straight line since $y = az + b$ is of the standard linear form. In effect y has been plotted against $\dfrac{1}{x^2}$.

Consider $\quad y = a\sqrt{x} + b$

Let $z = \sqrt{x}$ and, as previously, if we plot values of y against z (in effect $\sqrt{x}$) we will obtain a straight line since $y = az + b$ is of the standard linear form.

Consider $\quad y = \dfrac{a}{\sqrt{x}} + b$

This may also be written equivalently as $y = \dfrac{a}{x^{1/2}} + b$ or $y = ax^{-1/2} + b$.

Let $z = \dfrac{1}{\sqrt{x}}$ and, as previously, if we plot values of y against z $\left(\text{in effect } \dfrac{1}{\sqrt{x}} \right)$ we will obtain a straight line since $y = az + b$ is the standard linear form.

Exercise 11.2

1) The following readings were taken during a test:

R (ohms)	85	73.3	64	58.8	55.8
I (amperes)	2	3	5	8	12

R and I are thought to be connected by an equation of the form $R = \dfrac{a}{I} + b$.

Verify that this is so by plotting R (y-axis) against $\dfrac{1}{I}$ (x-axis) and hence find values for a and b.

2) In the theory of the moisture content of thermal insulation efficiency of porous materials the following table gives values of μ, the diffusion constant of the material, and k_m, the thermal conductivity of damp insulation material:

μ	1.3	2.7	3.8	5.4	7.2	10.0
k_m	0.0336	0.0245	0.0221	0.0203	0.0192	0.0183

Find the equation connecting μ and k_m if it is of the form $k_m = a + \dfrac{b}{\mu}$ where a and b are constants.

3) The accompanying table gives the corresponding values of the pressure, p, of mercury and the volume, V, of a given mass of gas at constant temperature.

p	90	100	130	150	170	190
v	16.66	13.64	11.54	9.95	8.82	7.89

By plotting p against the reciprocal of V obtain a relationship between p and V.

4) The approximate number of a type of bacteria, B, is checked regularly and recorded in the table below.

Bacteria, B ($\times 10^3$)	5	28.5	41.0	113.0	253.6	450.0
Time, t (hours)	1.0	2.5	3.0	5.0	7.5	10.0

It is thought that the growth is related according to the law $B = mt^2 + c$ where m and c are constants. By plotting a suitable graph verify this to be true and evaluate m and c.

5) In an experiment, the resistance R of copper wire of various diameters d mm was measured and the following readings were obtained.

d (mm)	0.1	0.2	0.3	0.4	0.5
R (ohms)	20	5	2.2	1.3	0.8

Show that $R = \dfrac{k}{d^2}$ and find a suitable value for k.

6) The following table gives the thickness T mm of a brass flange brazed to a copper pipe of internal diameter D mm:

T mm	15.5	17.8	19.5	20.9	22.2	23.3
D mm	50	100	150	200	250	300

Show that T and D are connected by an equation of the form $T = a\sqrt{D} + b$, find the values of constants a and b, and find the thickness of the flange for a 70 mm diameter pipe.

7) The table shows how the coefficient of friction, μ, between a belt and a pulley varies with the speed, v m/s, of the belt. By plotting a graph show that $\mu = m\sqrt{v} + c$ and find the values of constants m and c.

μ	0.26	0.29	0.32	0.35	0.38
v	2.22	5.00	8.89	13.89	20.00

8) Using the table below show that the values are in agreement with the law $y = \dfrac{m}{\sqrt{x}} + c$. Hence evaluate the constants m and c.

x	0.2	0.8	1.2	1.8	2.5	4.4
y	1.62	1.51	1.49	1.47	1.46	1.44

9) On a test bed a projectile, starting from rest over a constant measured distance, is subjected to different accelerations. The table below records these accelerations, A, and the times, t, taken to complete the distance.

A (ms^{-2})	5	10	15	20	25	30
t (s)	5.00	3.35	2.60	2.20	1.90	1.70

If the law connecting A and t is thought to be of the form $t = aA^{-1/2} + b$ plot a suitable straight line graph, and find the constants a and b.

REDUCING EQUATIONS OF THE TYPE $z = a \cdot t^n$ TO THEIR LOGARITHMIC FORM

Consider the following relationship, in which z and t are the variables whilst a and n are constants.

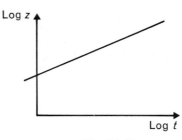

$$z = at^n$$

Taking logs we get:

$$\log z = \log(t^n a)$$
$$= \log t^n + \log a$$
$$\therefore \log z = n(\log t) + \log a$$

Fig. 11.13

The given values of the variables will satisfy this equation if they satisfy the original equation. Comparing this equation with $y = mx + c$, which is the standard equation of a straight line, we see that if we plot $\log z$ on the y-axis and $\log t$ on the x-axis, the result will be a straight line (Fig. 11.13).

The values of the constants n and a may be found using the *two-point method*.

The procedure is shown in Example 11.13 which follows.

EXAMPLE 11.3

The law connecting two quantities z and t is of the form $z = a \cdot t^n$ Find the values of the constants a and n given the following pairs of values:

z	3.170	4.603	7.499	10.50	15.17
t	7.980	9.863	13.03	15.81	19.50

By taking logs and rearranging (see text) we have

$$\log z = n \log t + \log a \qquad [1]$$

For the numerical part of the solution we may use common or natural logarithms. The solution given below uses common logs. You may find it instructive to work through the example using natural logs and verify that the same results are obtained.

From the given values, using logarithms to the base 10:

$\log_{10} z$	0.5011	0.6631	0.8750	1.0212	1.1810
$\log_{10} t$	0.9020	0.9940	1.1149	1.1990	1.2900

As usual we shall use the two-point method of finding the constants (Fig. 11.14).

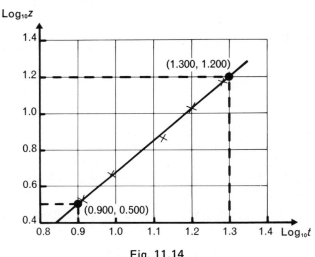

Fig. 11.14

Point (0.900, 0.500) lies on the line, and substituting in equation [1] gives

$$0.500 = n(0.900) + \log_{10} a \qquad [2]$$

Point (1.300, 1.200) lies on the line, and substituting in equation [1] gives

$$1.200 = n(1.300) + \log_{10} a \qquad [3]$$

Subtracting equation [2] from equation [3] gives

$$0.7 = 0.4n$$

∴ $$n = 1.75$$

Substituting in the equation [2] gives

$$0.500 = 1.75(0.900) + \log_{10} a$$

∴ $$\log_{10} a = -1.075$$

∴ $$a = 0.084$$

USE OF FULL LOGARITHMIC GRAPH PAPER

We have seen in Chapter 10 how logarithmic scales may be used on graph paper in place of the more usual linear scales. You will remember that logarithmic plots may be made without the necessity of finding the logarithms of each given value.

Log-log (or full logarithmic) graph paper, shown in Fig. 11.15, has logarithmic scales on both vertical and horizontal axes.

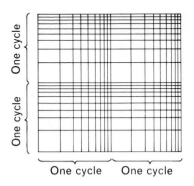

Fig. 11.15

The use of full logarithmic graph paper enables power relationships between two variables to be verified quickly.

A straight line graph on full logarithmic (or log–log) graph paper indicates a relationship between the variables x and y of the form $y = ax^n$, a and n being constants.

The use of these scales and the special graph paper is shown below by solving Example 11.3 already worked using ordinary linear paper.

EXAMPLE 11.4

The law connecting two quantities z and t is of the form $z = at^n$. Find the values of the constants a and n.

z	3.170	4.603	7.499	10.50	15.17
t	7.980	9.863	13.03	15.81	19.50

The relationship

$$z = at^n$$

gives, by taking logs of both sides,

$$\log z = n \log t + \log a \qquad [1]$$

Instead of looking up each $\log z$ and each $\log t$ individually we can plot the given values of z and t on log scales as shown in Fig. 11.16.

Choice of scales is largely governed by the log graph paper, the scales comprising a repeating pattern. Each cycle is ten times the previous one (e.g. 0.1 to 1, 1 to 10, 10 to 100, etc.).

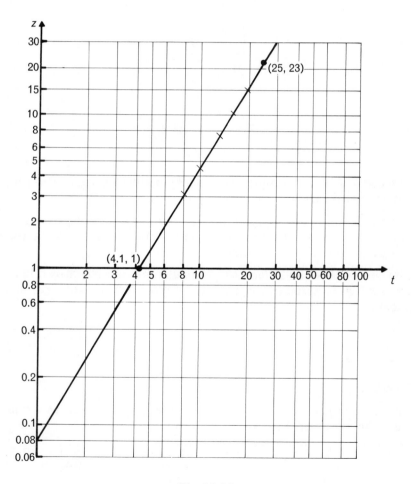

Fig. 11.16

The constants are again found by the two-point method:

Point $(25, 23)$ lies on the line, and putting these values in equation [1] gives

$$\log 23 \;=\; n(\log 25) + \log a \qquad\qquad [2]$$

Point $(4.1, 1)$ lies on the line, and putting these values in equation [1] gives

$$\log 1 \;=\; n(\log 4.1) + \log a \qquad\qquad [3]$$

Subtracting equation [3] from equation [2] gives

$$\log 23 - \log 1 \;=\; n(\log 25 - \log 4.1)$$

$$\therefore \qquad \log\left(\frac{23}{1}\right) \;=\; n\left\{\log\left(\frac{25}{4.1}\right)\right\}$$

or $\qquad\qquad \log 23 \;=\; n(\log 6.10)$

Here we again have to choose between natural or common logarithms. We will proceed using natural logarithms (logarithms to the base 'e').

Thus $\qquad\qquad n \;=\; \dfrac{\log_e 23}{\log_e 6.10} \;=\; \dfrac{3.14}{1.81} \;=\; 1.73$

Now to find a we will substitute $n = 1.73$ into equation [3]

$$\log_e 1 \;=\; 1.73(\log_e 4.1) + \log_e a$$

$$\therefore \qquad\qquad 0 \;=\; 1.73 \times 1.41 + \log_e a$$

$$\therefore \qquad \log_e a \;=\; -2.44$$

from which $\qquad a \;=\; 0.087$

Thus the required law is

$$z \;=\; 0.087t^{1.73}$$

This result verifies that obtained in Example 11.3, bearing in mind that the small difference in the figures obtained is due to inaccuracies caused by rounding to three significant figures during the solution.

This is acceptable since similar inaccuracies are present when selecting the 'best straight line' through the plotted points on the graph.

Exercise 11.3

1) The following values of x and y follow a law of the type $y = ax^n$. By plotting $\log y$ (vertically) against $\log x$ (horizontally) find values for a and n.

x	1	2	3	4	5
y	3	12	27	48	75

2) The following results were obtained in an experiment to find the relationships between the luminosity I of a metal filament lamp and the voltage V.

V	40	60	80	100	120
I	5.1	26.0	82	200	414

The law is thought to be of the type $I = aV^n$ Test this by plotting $\log I$ (vertically) against $\log V$ (horizontally) and find suitable values for a and n.

3) The relationship between power P (watts), the e.m.f. E (volts) and the resistance R (ohms) is thought to be of the form $P = \dfrac{E^n}{R}$. In an experiment in which R was kept constant the following results were obtained:

E	5	10	15	20	25	30
P	2.5	10	22.5	40	62.5	90

Verify the law and find the values of the constants n and R.

4) The following results were obtained in an experiment to find the relationship between the luminosity I of a metal filament lamp and the voltage V.

V	60	80	100	120	140
I	11	20.5	89	186	319

Allowing for the fact that an error was made in one of the readings show that the law between I and V is of the form $I = aV^n$ and find the probable correct value of the reading. Find the value of n.

5) Two quantities t and m are connected by a law of the type $t = am^b$ and the coordinates of two points which satisfy the equation are $(8, 6.8)$ and $(20, 26.9)$. Find the law.

6) The intensity of radiation, R, from certain radioactive materials at a particular time t is thought to follow the law $R = kt^n$. In an experiment to test this the following values were obtained:

R	58	43.5	26.5	14.5	10
t	1.5	2	3	5	7

Show that the assumption was correct and evaluate k and n.

MATRICES

After reaching the end of this chapter you should be able to:

1. *Recognise the notation of a matrix.*
2. *Calculate the sum and difference of two matrices (2 × 2 only).*
3. *Calculate the product of two 2 × 2 matrices.*
4. *Demonstrate that the product of two matrices is, in general, non-commutative.*
5. *Define the unit matrix.*
6. *Recognise the notation for a determinant.*
7. *Evaluate a 2 × 2 determinant.*
8. *Solve simultaneous linear equations with two unknowns using determinants.*
9. *Describe the meaning of a determinant whose value is zero and define a singular matrix.*
10. *Obtain the inverse of a 2 × 2 matrix.*
11. *Solve simultaneous linear equations with two unknowns by means of matrices.*

INTRODUCTION

The block within which a printer sets his type, and a car radiator, could each be called a *matrix*.

A matrix in mathematics is any rectangular array of numbers, usually enclosed in brackets.

Before using matrices in practical problems we first look at how they are ordered, added, subtracted and manipulated in other ways.

ELEMENT

Each number or symbol in a matrix is called an *element* of the matrix.

ORDER

The *dimension* or *order* of a matrix is stated by the number of rows followed by the number of columns in the rectangular array.

e.g.

Matrix	$\begin{pmatrix} 1 & 2 \\ 3 & 4 \end{pmatrix}$	$\begin{pmatrix} a & 2 & -3 \\ 4 & b & x \end{pmatrix}$	$\begin{pmatrix} \sin\theta & 1 \\ \cos\theta & 2 \\ \tan\theta & 3 \end{pmatrix}$	(6)
Order	2×2	2×3	3×2	1×1

EQUALITY

If two matrices are equal, then they must be of the same order and their corresponding elements must be equal.

Thus if $\begin{pmatrix} 2 & 3 & x \\ a & 5 & -2 \end{pmatrix} = \begin{pmatrix} 2 & 3 & 4 \\ -1 & 5 & -2 \end{pmatrix}$ then $x = 4$ and $a = -1$

ADDITION AND SUBTRACTION

Two matrices may be added or subtracted only if they are of the *same order*. We say the matrices are *conformable* for addition (or subtraction) and we add (or subtract) by combining corresponding elements.

EXAMPLE 12.1

If $A = \begin{pmatrix} 3 & 4 \\ 5 & 6 \end{pmatrix}$ and $B = \begin{pmatrix} 0 & 6 \\ 5 & 2 \end{pmatrix}$ determine: **a)** $C = A + B$
and   **b)** $D = A - B$

a) $C = \begin{pmatrix} 3 & 4 \\ 5 & 6 \end{pmatrix} + \begin{pmatrix} 0 & 6 \\ 5 & 2 \end{pmatrix} = \begin{pmatrix} 3+0 & 4+6 \\ 5+5 & 6+2 \end{pmatrix} = \begin{pmatrix} 3 & 10 \\ 10 & 8 \end{pmatrix}$

b) $D = \begin{pmatrix} 3 & 4 \\ 5 & 6 \end{pmatrix} - \begin{pmatrix} 0 & 6 \\ 5 & 2 \end{pmatrix} = \begin{pmatrix} 3-0 & 4-6 \\ 5-5 & 6-2 \end{pmatrix} = \begin{pmatrix} 3 & -2 \\ 0 & 4 \end{pmatrix}$

ZERO OR NULL MATRIX

A *zero* or *null* matrix, denoted by O, is one in which all the elements are zero. It may be of any order.

Thus $\begin{pmatrix} 0 & 0 \\ 0 & 0 \end{pmatrix}$ is a zero matrix of order 2. It behaves like zero in the real number system.

IDENTITY OR UNIT MATRIX

The *identity* matrix can be of any suitable order with all the main diagonal elements 1 and the remaining elements 0. It is denoted by I and behaves like unity in the real number system.

Thus $\begin{pmatrix} 1 & 0 \\ 0 & 1 \end{pmatrix}$ is a unit matrix of order 2

TRANSPOSE

The *transpose* of a matrix A is written as A' or A^T. When the row of a matrix is interchanged with its corresponding column, that is row 1 becomes column 1 and row 2 becomes column 2 and so on, then the matrix is transposed.

Thus if $A = \begin{pmatrix} 1 & 2 & -3 \\ 4 & 7 & 0 \end{pmatrix}$ then $A' = \begin{pmatrix} 1 & 4 \\ 2 & 7 \\ -3 & 0 \end{pmatrix}$

Exercise 12.1

1) State the order of each of the following matrices:

(a) $\begin{pmatrix} 1 & 2 \\ 3 & 4 \end{pmatrix}$ (b) $\begin{pmatrix} 5 \\ -6 \end{pmatrix}$ (c) $\begin{pmatrix} a & b & 4 \\ 2 & 3 & 5 \\ x & -6 & 0 \end{pmatrix}$

(d) $\begin{pmatrix} 1 & -2 & -3 & -4 \\ 6 & 2 & 0 & -1 \end{pmatrix}$

2) How many elements are there in:

(a) a 3×3 matrix   (b) a 2×2 matrix
(c) a square matrix of order n?

3) Write down the transpose of each matrix in Question 1).

4) Combine the following matrices:

(a) $\begin{pmatrix} 2 & 1 \\ 3 & 2 \end{pmatrix} + \begin{pmatrix} -2 & -1 \\ 6 & 0 \end{pmatrix}$ (b) $\begin{pmatrix} 2 & 1 \\ 3 & 2 \end{pmatrix} - \begin{pmatrix} -2 & -1 \\ 6 & 0 \end{pmatrix}$

(c) $\begin{pmatrix} \frac{1}{2} & 1 \\ \frac{1}{3} & \frac{1}{5} \end{pmatrix} + \begin{pmatrix} \frac{1}{3} & -\frac{1}{2} \\ \frac{1}{2} & \frac{4}{5} \end{pmatrix}$

5) Determine a, b and c if $(a \quad b \quad c) - (-3 \quad 4 \quad 1) = (-5 \quad 1 \quad 0)$.

6) Complete $\begin{pmatrix} \frac{1}{2} & \frac{1}{4} \\ \frac{1}{5} & \frac{1}{6} \end{pmatrix} - \begin{pmatrix} \frac{1}{6} & \frac{1}{5} \\ \frac{1}{6} & \frac{1}{9} \end{pmatrix}$

7) Solve the equation $X - \begin{pmatrix} 1 & 3 \\ 5 & -2 \end{pmatrix} = \begin{pmatrix} 4 & 5 \\ 7 & 0 \end{pmatrix}$ where X is a 2×2 matrix.

8) If $\begin{pmatrix} 4 \\ 5 \end{pmatrix} + \begin{pmatrix} x \\ y \end{pmatrix} = \begin{pmatrix} 4 \\ 10 \end{pmatrix}$, determine $\begin{pmatrix} x \\ y \end{pmatrix}$

MULTIPLICATION OF A MATRIX BY A REAL NUMBER

A matrix may be multiplied by a number in the following way:

$$4\begin{pmatrix} 2 & 3 \\ 7 & -1 \end{pmatrix} = \begin{pmatrix} 4 \times 2 & 4 \times 3 \\ 4 \times 7 & 4 \times (-1) \end{pmatrix} = \begin{pmatrix} 8 & 12 \\ 28 & -4 \end{pmatrix}$$

Conversely the common factor of each element in a matrix may be written outside the matrix. Thus $\begin{pmatrix} 9 & 3 \\ 42 & 15 \end{pmatrix} = 3\begin{pmatrix} 3 & 1 \\ 14 & 5 \end{pmatrix}$

MATRIX MULTIPLICATION

Two matrices can only be multiplied together if the number of columns in the first matrix is equal to the number of rows in the second matrix. We say that the matrices are *conformable* for multiplication. The method for multiplying together a pair of 2×2 matrices is as follows

$$\begin{pmatrix} a & b \\ c & d \end{pmatrix} \times \begin{pmatrix} e & f \\ g & h \end{pmatrix} = \begin{pmatrix} ae + bg & af + bh \\ ce + dg & cf + dh \end{pmatrix}$$

EXAMPLE 12.2

a) $\begin{pmatrix} 2 & 3 \\ 4 & 5 \end{pmatrix} \times \begin{pmatrix} 7 & 1 \\ 0 & 6 \end{pmatrix} = \begin{pmatrix} (2 \times 7) + (3 \times 0) & (2 \times 1) + (3 \times 6) \\ (4 \times 7) + (5 \times 0) & (4 \times 1) + (5 \times 6) \end{pmatrix}$

$$= \begin{pmatrix} 14 & 20 \\ 28 & 34 \end{pmatrix}$$

b) $\begin{pmatrix} 3 & 4 \\ 2 & 5 \end{pmatrix} \times \begin{pmatrix} 6 \\ 7 \end{pmatrix} = \begin{pmatrix} (3 \times 6) + (4 \times 7) \\ (2 \times 6) + (5 \times 7) \end{pmatrix} = \begin{pmatrix} 46 \\ 47 \end{pmatrix}$

c) $\begin{pmatrix} 3 \\ 2 \end{pmatrix} \times \begin{pmatrix} 4 & 6 \\ 5 & 7 \end{pmatrix}$ This is not possible since the matrices are not comformable.

EXAMPLE 12.3

Form the products AB and BA given that $A = \begin{pmatrix} 1 & 2 \\ 3 & 4 \end{pmatrix}$ and

$B = \begin{pmatrix} 5 & 6 \\ 7 & 8 \end{pmatrix}$ and hence show that $AB \neq BA$.

$$AB = \begin{pmatrix} 1 & 2 \\ 3 & 4 \end{pmatrix} \begin{pmatrix} 5 & 6 \\ 7 & 8 \end{pmatrix} = \begin{pmatrix} (1 \times 5) + (2 \times 7) & (1 \times 6) + (2 \times 8) \\ (3 \times 5) + (4 \times 7) & (3 \times 6) + (4 \times 8) \end{pmatrix}$$

$$= \begin{pmatrix} 19 & 22 \\ 43 & 50 \end{pmatrix}$$

$$BA = \begin{pmatrix} 5 & 6 \\ 7 & 8 \end{pmatrix} \begin{pmatrix} 1 & 2 \\ 3 & 4 \end{pmatrix} = \begin{pmatrix} (5 \times 1) + (6 \times 3) & (5 \times 2) + (6 \times 4) \\ (7 \times 1) + (8 \times 3) & (7 \times 2) + (8 \times 4) \end{pmatrix}$$

$$= \begin{pmatrix} 23 & 34 \\ 31 & 46 \end{pmatrix}$$

As we see the results are different and, in general, matrix multiplication is non-commutative, i.e. $AB \neq BA$

Exercise 12.2

1) If $A = \begin{pmatrix} 3 & 0 \\ -2 & 1 \end{pmatrix}$ and $B = \begin{pmatrix} -4 & 1 \\ 3 & -2 \end{pmatrix}$ determine:

(a) $2A$ (b) $3B$ (c) $2A + 3B$ (d) $2A - 3B$

2) Calculate the following products:

(a) $\begin{pmatrix} 3 & 1 \\ 2 & 0 \end{pmatrix} \begin{pmatrix} 4 & -1 \\ 2 & 3 \end{pmatrix}$ (b) $\begin{pmatrix} 2 & 1 \\ 3 & 1 \end{pmatrix} \begin{pmatrix} 1 & 0 \\ 0 & 1 \end{pmatrix}$ (c) $\begin{pmatrix} 2 & 1 \\ 4 & 2 \end{pmatrix} \begin{pmatrix} 2 & 3 \\ 1 & 5 \end{pmatrix}$

(d) $\begin{pmatrix} 1 & 0 \\ 0 & 1 \end{pmatrix} \begin{pmatrix} a & b \\ c & d \end{pmatrix}$ (e) $\begin{pmatrix} k & 0 \\ 0 & k \end{pmatrix} \begin{pmatrix} a & b \\ c & d \end{pmatrix}$

3) If $A = \begin{pmatrix} 1 & 2 \\ 3 & 4 \end{pmatrix}$ and $B = \begin{pmatrix} 2 & -1 \\ 1 & 3 \end{pmatrix}$ calculate:

(a) A^2 (that is $A \times A$)   (b) B^2 (c) $2AB$

(d) $A^2 + B^2 + 2AB$ (e) $(A + B)^2$

DETERMINANT OF A SQUARE MATRIX OF ORDER 2

If matrix $A = \begin{pmatrix} a & b \\ c & d \end{pmatrix}$ then its *determinant* is denoted by $|A|$ or $\det A$ and the result is a *number* given by

$$|A| = \begin{vmatrix} a & b \\ c & d \end{vmatrix} = ad - bc$$

EXAMPLE 12.4

Evaluate $|A|$ if $A = \begin{pmatrix} 1 & -2 \\ 3 & 4 \end{pmatrix}$.

$$|A| = \begin{vmatrix} 1 & -2 \\ 3 & 4 \end{vmatrix} = 1 \times 4 - (-2) \times 3 = 10$$

SOLUTION OF SIMULTANEOUS LINEAR EQUATIONS USING DETERMINANTS

To solve simultaneous linear equations with two unknowns using determinants, the following procedure is used.

(a) Write out the two equations in order:

$$a_1 x + b_1 y = c_1$$
$$a_2 x + b_2 y = c_2$$

(b) Calculate $\Delta = \begin{vmatrix} a_1 & b_1 \\ a_2 & b_2 \end{vmatrix}$

(c) Then $x = \dfrac{\begin{vmatrix} c_1 & b_1 \\ c_2 & b_2 \end{vmatrix}}{\Delta}$ and $y = \dfrac{\begin{vmatrix} a_1 & c_1 \\ a_2 & c_2 \end{vmatrix}}{\Delta}$

EXAMPLE 12.5

By using determinants, solve the simultaneous equations

$$3x + 4y = 22$$
$$2x + 5y = 24$$

Now $\Delta = \begin{vmatrix} 3 & 4 \\ 2 & 5 \end{vmatrix} = (3 \times 5) - (4 \times 2) = 7$

Thus $x = \dfrac{\begin{vmatrix} 22 & 4 \\ 24 & 5 \end{vmatrix}}{7} = \dfrac{(22 \times 5) - (4 \times 24)}{7} = \dfrac{14}{7} = 2$

And $y = \dfrac{\begin{vmatrix} 3 & 22 \\ 2 & 24 \end{vmatrix}}{7} = \dfrac{(3 \times 24) - (22 \times 2)}{7} = \dfrac{28}{7} = 4$

Exercise 12.3

1) Evaluate the following determinants:

(a) $\begin{vmatrix} 5 & 2 \\ 3 & 6 \end{vmatrix}$ (b) $\begin{vmatrix} 7 & 4 \\ 5 & 2 \end{vmatrix}$ (c) $\begin{vmatrix} 6 & 8 \\ 2 & 5 \end{vmatrix}$

2) Solve the following simultaneous equations by using determinants:

(a) $3x + 4y = 11$ (b) $5x + 3y = 29$ (c) $4x - 6y = -2.5$
$\quad\;\; x + 7y = 15$ $\quad\;\; 4x + 7y = 37$ $\quad\;\; 7x - 5y = -0.25$

THE INVERSE OF A SQUARE MATRIX OF ORDER 2

Instead of dividing a number by 5 we can multiply by $\frac{1}{5}$ and obtain the same result.

Thus $\frac{1}{5}$ is the multiplicative inverse of 5. That is $5 \times \frac{1}{5} = 1$

In matrix algebra we never divide by a matrix but multiply instead by the inverse. The inverse of matrix A is denoted by A^{-1} and is such that

$$AA^{-1} = \begin{pmatrix} 1 & 0 \\ 0 & 1 \end{pmatrix} = I, \quad \text{the identity matrix.}$$

To find the inverse, A^{-1}, of the square matrix $A = \begin{pmatrix} a & b \\ c & d \end{pmatrix}$

we use the expression: $A^{-1} = \dfrac{1}{|A|} \begin{pmatrix} d & -b \\ -c & a \end{pmatrix} = \dfrac{1}{ad - bc} \begin{pmatrix} d & -b \\ -c & a \end{pmatrix}$

EXAMPLE 12.6

Determine the inverse of $A = \begin{pmatrix} 1 & -2 \\ 3 & 4 \end{pmatrix}$ and verify the result.

Now $|A| = \begin{vmatrix} 1 & -2 \\ 3 & 4 \end{vmatrix} = (1 \times 4) - (3 \times -2) = 10$

Hence $A^{-1} = \frac{1}{10} \begin{pmatrix} 4 & 2 \\ -3 & 1 \end{pmatrix} = \begin{pmatrix} 0.4 & 0.2 \\ -0.3 & 0.1 \end{pmatrix}$

To verify the result we have

$$AA^{-1} = \begin{pmatrix} 1 & -2 \\ 3 & 4 \end{pmatrix} \begin{pmatrix} 0.4 & 0.2 \\ -0.3 & 0.1 \end{pmatrix} = \begin{pmatrix} 1 & 0 \\ 0 & 1 \end{pmatrix} = I$$

SINGULAR MATRIX

A matrix which does not have an inverse is called a *singular matrix*.
This happens when $|A| = 0$

For example, since $\begin{vmatrix} 3 & 6 \\ 1 & 2 \end{vmatrix} = (3 \times 2) - (6 \times 1) = 0$

then $\begin{pmatrix} 3 & 6 \\ 1 & 2 \end{pmatrix}$ is a singular matrix.

Exercise 12.4

Decide whether each of the matrices in Question 1–9 has an inverse. If the inverse exists, find it.

1) $\begin{pmatrix} 2 & 5 \\ 1 & 4 \end{pmatrix}$ 2) $\begin{pmatrix} 2 & 5 \\ 1 & 3 \end{pmatrix}$ 3) $\begin{pmatrix} 3 & 2 \\ 1 & 2 \end{pmatrix}$ 4) $\begin{pmatrix} 4 & 10 \\ 2 & 5 \end{pmatrix}$

5) $\begin{pmatrix} 224 & 24 \\ 24 & 4 \end{pmatrix}$ 6) $\begin{pmatrix} a & -b \\ -a & b \end{pmatrix}$ 7) $\begin{pmatrix} 2 & 3 \\ -1 & 1 \end{pmatrix}$ 8) $\begin{pmatrix} 2 & -3 \\ 1 & 5 \end{pmatrix}$

9) $\begin{pmatrix} 1 & 1 \\ 0 & 1 \end{pmatrix}$

10) Given that $A = \begin{pmatrix} 1 & 0 \\ 3 & 2 \end{pmatrix}$ and $B = \begin{pmatrix} 3 & 5 \\ 1 & 2 \end{pmatrix}$, calculate:

(a) A^{-1} (b) B^{-1} (c) $B^{-1}A^{-1}$ (d) AB

(e) $(AB)^{-1}$ (f) Compare the answers to (c) and (e)

SYSTEMS OF LINEAR EQUATIONS

Given the system of equations $\left.\begin{aligned} 5x + y &= 7 \\ 3x - 4y &= 18 \end{aligned}\right\}$

we can rewrite it in the form $\begin{pmatrix} 5x + y \\ 3x - 4y \end{pmatrix} = \begin{pmatrix} 7 \\ 18 \end{pmatrix}$

or $\begin{pmatrix} 5 & 1 \\ 3 & -4 \end{pmatrix}\begin{pmatrix} x \\ y \end{pmatrix} = \begin{pmatrix} 7 \\ 18 \end{pmatrix}$

That is $\begin{pmatrix}\text{Matrix of}\\\text{coefficients}\end{pmatrix}\begin{pmatrix}\text{Matrix of}\\\text{variables}\end{pmatrix} = \begin{pmatrix}\text{Matrix of}\\\text{constants}\end{pmatrix}$

Denote the matrix of coefficients by C and its inverse by C^{-1}.

Then
$$|C| = \begin{vmatrix} 5 & 1 \\ 3 & -4 \end{vmatrix} = 5\times(-4)-1\times3 = -23$$

and
$$C^{-1} = \frac{1}{-23}\begin{pmatrix} -4 & -1 \\ -3 & 5 \end{pmatrix} = \frac{1}{23}\begin{pmatrix} 4 & 1 \\ 3 & -5 \end{pmatrix}$$

Now
$$C\begin{pmatrix} x \\ y \end{pmatrix} = \begin{pmatrix} 7 \\ 18 \end{pmatrix}$$

and multiplying both sides by C^{-1} gives

$$C^{-1}C\begin{pmatrix} x \\ y \end{pmatrix} = C^{-1}\begin{pmatrix} 7 \\ 18 \end{pmatrix}$$

$\therefore$
$$I\begin{pmatrix} x \\ y \end{pmatrix} = C^{-1}\begin{pmatrix} 7 \\ 18 \end{pmatrix}$$

or
$$\begin{pmatrix} 1 & 0 \\ 0 & 1 \end{pmatrix}\times\begin{pmatrix} x \\ y \end{pmatrix} = \frac{1}{23}\begin{pmatrix} 4 & 1 \\ 3 & -5 \end{pmatrix}\times\begin{pmatrix} 7 \\ 18 \end{pmatrix}$$

$\therefore$
$$\begin{pmatrix} 1\times x & 0\times y \\ 0\times x & 1\times y \end{pmatrix} = \frac{1}{23}\begin{pmatrix} 4\times7+ & 1\times18 \\ 3\times7+(-5)\times18 \end{pmatrix}$$

$\therefore$
$$\begin{pmatrix} x \\ y \end{pmatrix} = \frac{1}{23}\begin{pmatrix} 46 \\ -69 \end{pmatrix}$$

$\therefore$
$$\begin{pmatrix} x \\ y \end{pmatrix} = \begin{pmatrix} 2 \\ -3 \end{pmatrix}$$

Thus comparing the matrices shows that $x = 2$ and $y = -3$

We would not normally perform multiplication by the unit matrix.

We did so here to illustrate that when a matrix, here $\begin{pmatrix} x \\ y \end{pmatrix}$, is multiplied by the unit matrix then it is unaltered.

This confirms that the unit matrix performs as unity (the number one) in normal arithmetic.

Exercise 12.5

Use matrix methods to solve each of the following systems of equations.

1) $\left.\begin{array}{l} x + y = 1 \\ 3x + 2y = 8 \end{array}\right\}$

2) $\left.\begin{array}{l} x + y = 6 \\ 3x - 2y = -7 \end{array}\right\}$

3) $\left.\begin{array}{l} 5x - 2y = 17 \\ 2x + 3y = 3 \end{array}\right\}$

4) $\left.\begin{array}{l} 3x - 2y = 12 \\ 4x + y = 5 \end{array}\right\}$

5) $\left.\begin{array}{l} 3x + 2y = 6 \\ 4x - y = 5 \end{array}\right\}$

6) $\left.\begin{array}{l} 3x - 4y = 26 \\ 5x + 6y = -20 \end{array}\right\}$

13.

COMPLEX NUMBERS

On reaching the end of this chapter you should be able to:

1. *Understand the necessity of extending the number system to include the square roots of negative numbers.*
2. *Define j as $\sqrt{-1}$.*
3. *Define a complex number as consisting of a real part and an imaginary part.*
4. *Define a complex number z in the algebraic form x + jy.*
5. *Determine the complex roots of $ax^2 + bx + c = 0$ when $b^2 < 4ac$ using the quadratic formula.*
6. *Perform the addition and subtraction of complex numbers in algebraic form.*
7. *Define the conjugate of a complex number in algebraic form.*
8. *Perform the multiplication and division of complex numbers in algebraic form.*
9. *Represent the algebraic form of a complex number on an Argand diagram, and show how it may be represented as a phasor.*

10. *Deduce that j may be considered to be an operator, such that when the phasor representing x + jy is multiplied by j it rotates the phasor through 90° anti-clockwise.*
11. *Understand how phasors on an Argand diagram may be added and subtracted in a manner similar to the addition and subtraction of vectors.*
12. *Show that the full polar form of a complex number is (cos θ + j sin θ) which may be abbreviated to r ∠θ.*
13. *Perform the operations involved in the conversion of complex numbers in algebraic form to polar form and vice versa.*
14. *Multiply and divide numbers in polar form.*
15. *Determine the square roots of a complex number.*
16. *Apply the above to problems arising from relevant engineering technology.*

INTRODUCTION

The solution of the quadratic equation $ax^2 + bx + c = 0$ is given by the formula

$$x = \frac{-b \pm \sqrt{b^2 - 4ac}}{2a}$$

When we use this formula most of the quadratic equations we meet, when solving engineering problems, are found to have roots which are ordinary positive or negative numbers.

Consider now the equation $x^2 - 4x + 13 = 0$.

Then
$$x = \frac{-(-4) \pm \sqrt{(-4)^2 - 4 \times 1 \times 13}}{2 \times 1}$$

$$= \frac{4 \pm \sqrt{16 - 52}}{2}$$

$$= \frac{4 \pm \sqrt{-36}}{2}$$

$$= \frac{4 \pm \sqrt{(-1)(36)}}{2}$$

$$= \frac{4 \pm \sqrt{(-1)} \times \sqrt{(36)}}{2}$$

$$= \frac{4}{2} \pm \sqrt{-1} \times \frac{6}{2}$$

$$= 2 \pm \sqrt{-1} \times 3$$

It is not possible to find the value of the square root of a negative number.

In order to try to find a meaning for roots of this type we represent $\sqrt{-1}$ by the symbol j.

(Books on pure mathematics often use the symbol i, but in engineering j is preferred as i is used for the instantaneous value of a current.)

Thus the roots of the above equation become $2 + j3$ and $2 - j3$.

DEFINITIONS

Expressions such as $2 + j3$ are called *complex numbers*. The number 2 is called the *real part* and j3 is called the *imaginary part*.

The general expression for a complex number is $x + jy$, which has a real part equal to x and an imaginary part equal to jy. The form $x + jy$ is said to be the *algebraic form* of a complex number. It may also be called the *cartesian form* or *rectangular notation*.

POWERS OF j

We have defined j such that

$$j = \sqrt{-1}$$

$\therefore$ squaring both sides of the equation gives

$$j^2 = (\sqrt{-1})^2 = -1$$

Hence	$j^3 = j^2 \times j = (-1) \times j = -j$
and	$j^4 = (j^2)^2 = (-1)^2 = 1$
and	$j^5 = j^4 \times j = 1 \times j = j$
and	$j^6 = (j^2)^3 = (-1)^3 = -1$

and so on.

The most used of the above relationships is $j^2 = -1$.

ADDITION AND SUBTRACTION OF COMPLEX NUMBERS IN ALGEBRAIC FORM

The real and imaginary parts must be treated separately. The real parts may be added and subtracted and also the imaginary parts may be added and subtracted, both obeying the ordinary laws of algebra.

Thus

$$(3 + j2) + (5 + j6) = 3 + j2 + 5 + j6$$
$$= (3 + 5) + j(2 + 6)$$
$$= 8 + j8$$

and

$$(1 - j2) - (-4 + j) = 1 - j2 + 4 - j$$
$$= (1 + 4) - j(2 + 1)$$
$$= 5 - j3$$

EXAMPLE 13.1

If z_1, z_2 and z_3 represent three complex numbers such that $z_1 = 1.6 + j2.3$, $z_2 = 4.3 - j0.6$ and $z_3 = -1.1 - j0.9$ find the complex numbers which represent:

a) $z_1 + z_2 + z_3$,

b) $z_1 - z_2 - z_3$.

a) $z_1 + z_2 + z_3 = (1.6 + j2.3) + (4.3 - j0.6) + (-1.1 - j0.9)$
$$= 1.6 + j2.3 + 4.3 - j0.6 - 1.1 - j0.9$$
$$= (1.6 + 4.3 - 1.1) + j(2.3 - 0.6 - 0.9)$$
$$= 4.8 + j0.8$$

b) $z_1 - z_2 - z_3$ = $(1.6 + j2.3) - (4.3 - j0.6) - (-1.1 - j0.9)$

$\qquad\qquad$ = $1.6 + j2.3 - 4.3 + j0.6 + 1.1 + j0.9$

$\qquad\qquad$ = $(1.6 - 4.3 + 1.1) + j(2.3 + 0.6 + 0.9)$

$\qquad\qquad$ = $-1.6 + j3.8$

MULTIPLICATION OF COMPLEX NUMBERS IN ALGEBRAIC FORM

Consider the product of two complex numbers, $(3 + j2)(4 + j)$.

The brackets are treated in exactly the same way as the rules of algebra, such that

$$(a + b)(c + d) = ac + bc + ad + bd$$

Hence $(3 + j2)(4 + j)$ = $3 \times 4 + j2 \times 4 + 3 \times j + j2 \times j$

$\qquad\qquad$ = $12 + j8 + j3 + j^2 2$

$\qquad\qquad$ = $12 + j8 + j3 - 2 \qquad$ since $\quad j^2 = -1$

$\qquad\qquad$ = $(12 - 2) + j(8 + 3)$

$\qquad\qquad$ = $10 + j11$

EXAMPLE 13.2

Express the product of $2 + j$, $-3 + j2$, and $1 - j$ as a single complex number.

Then $(2 + j)(-3 + j2)(1 - j)$ = $(2 + j)(-3 + j2 + j3 - j^2 2)$

$\qquad\qquad$ = $(2 + j)(-1 + j5)$

$\qquad\qquad\qquad\qquad$ since $\quad j^2 = -1$

$\qquad\qquad$ = $-2 - j + j10 + j^2 5$

$\qquad\qquad$ = $-7 + j9 \qquad$ since $\quad j^2 = -1$

CONJUGATE COMPLEX NUMBERS

Consider $(x + jy)(x - jy)$ = $x^2 + jxy - jxy - j^2 y$

$\qquad\qquad$ = $x^2 - (-1)y^2$

$\qquad\qquad$ = $x^2 + y^2$

Hence we have the product of two complex numbers which produces a real number since it does not have a j term. If $x + jy$ represents a complex number then $x - jy$ is known as its *conjugate* (and vice versa). For example the conjugate of $(3 + j4)$ is $(3 - j4)$ and their product is

$$(3 + j4)(3 - j4) = 9 + j12 - j12 - j^2 16 = 9 - (-1)16$$

$$= 25 \quad \text{which is a real number}$$

DIVISION OF COMPLEX NUMBERS IN ALGEBRAIC FORM

Consider $\dfrac{(4 + j5)}{(1 - j)}$. We use the method of rationalising the denominator.

This means removing the j terms from the bottom line of the fraction. If we multiply $(1 - j)$ by its conjugate $(1 + j)$ the result will be a real number. Hence, in order not to alter the value of the given expression, we will multiply both the numerator and the denominator by $(1 + j)$.

Thus
$$\frac{(4 + 5j)}{(1 - j)} = \frac{(4 + j5)(1 + j)}{(1 - j)(1 + j)}$$

$$= \frac{4 + j5 + j4 + j^2 5}{1 - j + j - j^2}$$

$$= \frac{4 + j9 + (-1)5}{1 - (-1)}$$

$$= \frac{-1 + j9}{2}$$

$$= -\frac{1}{2} + j\frac{9}{2}$$

$$= -0.5 + j4.5$$

EXAMPLE 13.3

The impedance Z of a circuit having a resistance and inductive reactance in series is given by the complex number $Z = 5 + j6$.

Find the admittance Y of the circuit if $Y = \dfrac{1}{Z}$.

Now
$$Y = \frac{1}{Z} = \frac{1}{5+j6}$$

The conjugate of the denominator is $5-j6$ and we therefore multiply both the numerator and denominator by $5-j6$.

Then

$$Y = \frac{(5-j6)}{(5+j6)(5-j6)}$$

$$= \frac{5-j6}{25+j30-j30-j^2 36}$$

$$= \frac{5-j6}{25-(-1)36} = \frac{5-j6}{61} = \frac{5}{61} - j\frac{6}{61} = 0.082 - j0.098$$

EXAMPLE 13.4

Two impedances Z_1 and Z_2 are given by the complex numbers $Z_1 = 1 + j5$ and $Z_2 = j7$. Find the equivalent impedance Z if

a) $Z = Z_1 + Z_2$ where Z_1 and Z_2 are in series,

b) $\dfrac{1}{Z} = \dfrac{1}{Z_1} + \dfrac{1}{Z_2}$ when Z_1 and Z_2 are in parallel.

a)
$$Z = Z_1 + Z_2 = (1+j5) + j7$$
$$= 1 + j5 + j7$$
$$= 1 + j12$$

b)
$$\frac{1}{Z} = \frac{1}{Z_1} + \frac{1}{Z_2} = \frac{1}{(1+j5)} + \frac{1}{j7}$$
$$= \frac{j7 + (1+j5)}{(1+j5)j7}$$
$$= \frac{1+j12}{j7+j^2 35}$$
$$= \frac{1+j12}{j7+(-1)35}$$

Thus
$$Z = \frac{j7 - 35}{1 + j12}$$

$$= \frac{(j7 - 35)(1 - j12)}{(1 + j12)(1 - j12)}$$

$$= \frac{j7 - 35 - j^2 84 + j420}{1 + j12 - j12 - j^2 144}$$

$$= \frac{j427 - 35 - (-1)84}{1 - (-1)144}$$

$$= \frac{49 + j427}{145}$$

$$= 0.338 + j2.945$$

Exercise 13.1

1) Add the following complex numbers:
(a) $3 + j5$, $7 + j3$,   and $8 + j2$,
(b) $2 - j7$, $3 + j8$,   and $-5 - j2$,
(c) $4 - j2$, $7 + j3$, $-5 - j6$,   and $2 - j5$.

2) Subtract the following complex numbers:
(a) $3 + j5$ from $2 + j8$,
(b) $7 - j6$ from $3 - j9$,
(c) $-3 - j5$ from $7 - j8$.

3) Simplify the following expressions giving the answers in the form $x + jy$:

(a) $(3 + j3)(2 + j5)$,

(b) $(2 - j6)(3 - j7)$,

(c) $(4 + j5)^2$,

(d) $(5 + j3)(5 - j3)$,

(e) $(-5 - j2)(5 + j2)$,

(f) $(3 - j5)(3 - j3)(1 + j)$,

(g) $\dfrac{1}{2 + j5}$,

(h) $\dfrac{2 + j5}{2 - j5}$,

(i) $\dfrac{-2 - j3}{5 - j2}$,

(j) $\dfrac{7 + j3}{8 - j3}$,

(k) $\dfrac{(1 + j2)(2 - j)}{(1 + j)}$,

(l) $\dfrac{4 + j2}{(2 + j)(1 - j3)}$

4) Find the real and imaginary parts of:

(a) $1 + \dfrac{j}{2}$

(b) $j3 + \dfrac{2}{j3}$

(c) $(j2)^2 + 3(j)^5 - j(j)$

5) Solve the following equations giving the answers in the form $x + jy$:

(a) $x^2 + 2x + 2 = 0$

(b) $x^2 + 9 = 0$

6) Find the admittance Y of a circuit if $Y = \dfrac{1}{Z}$ where $Z = 1.3 + j0.6$

7) Three impedances Z_1, Z_2, and Z_3 are represented by the complex numbers $Z_1 = 2 + j$, $Z_2 = 1 + j$, and $Z_3 = j2$. Find the equivalent impedance Z if:

(a) $Z = Z_1 + Z_2 + Z_3$,

(b) $\dfrac{1}{Z} = \dfrac{1}{Z_1} + \dfrac{1}{Z_2} + \dfrac{1}{Z_3}$,

(c) $Z = \dfrac{1}{\dfrac{1}{Z_1} + \dfrac{1}{Z_2}} + Z_3.$

THE ARGAND DIAGRAM

When plotting a graph, cartesian coordinates are generally used to plot the points. Thus the position of the point P (Fig. 13.1) is defined by the coordinates $(3, 2)$ meaning that $x = 3$ and $y = 2$.

Complex numbers may be represented in a similar way on the Argand diagram. The real part of the complex number is plotted along the horizontal real-axis whilst the imaginary part is plotted along the vertical imaginary, or j-axis.

However a complex number is denoted, not by a point but, as a *phasor*. A phasor is a line where regard is paid both to its magnitude and to its direction. Hence in Fig. 13.2 the complex number $4 + j3$ is represented by the phasor $\overrightarrow{OQ}$, the end Q of the line being found by plotting 4 units along the real-axis and 3 units along the j-axis.

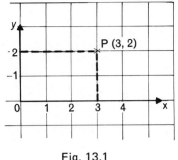
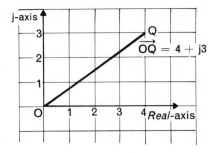

Fig. 13.1   Fig. 13.2

A single letter, the favourite being z, is often used to denote a phasor which represents a complex number. Thus if $z = x + jy$ it is understood that z represents a phasor and not a simple numerical value.

Four typical complex numbers z_1, z_2, z_3, and z_4 are shown on the Argand diagram in Fig. 13.3.

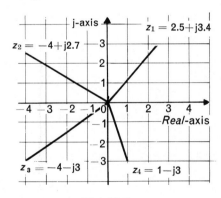

Fig. 13.3

A real number such as 2.7 may be regarded as a complex number with a zero imaginary part, i.e. $2.7 + j0$, and may be represented on the Argand diagram (Fig. 13.4) as the phasor $z = 2.7$ denoted by $\overrightarrow{OA}$ in the diagram.

A number such as j3 is said to be wholly imaginary and may be regarded as a complex number having a zero real part, i.e. $0 + j3$, and may be represented on the Argand diagram (Fig. 13.4) as the phasor $z = j3$ denoted by $\overrightarrow{OB}$ in the diagram.

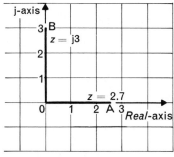

Fig. 13.4

THE j-OPERATOR

Consider the real number 3 shown on the Argand diagram, in Fig. 13.5.

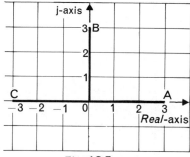

Fig. 13.5

It may be denoted by $\overrightarrow{OA}$. (This is a phasor because it has magnitude and direction.)

If we now multiply the real number 3 by j we obtain the complex number j3 which may be represented by the phasor $\overrightarrow{OB}$.

It follows that the effect of j on phasor $\overrightarrow{OA}$ is to make it become phasor $\overrightarrow{OB}$,

that is $$\overrightarrow{OB} = j\overrightarrow{OA}$$

Hence j is known as an operator (called the 'j-operator') which, when applied to a phasor, alters its direction by 90° in an anti-clockwise direction without changing its magnitude.

If we now operate on the phasor $\overrightarrow{OB}$ we shall obtain, therefore, phasor $\overrightarrow{OC}$.

In equation form this is

$$\overrightarrow{OC} = j\overrightarrow{OB}$$

but since $\overrightarrow{OB} = j\overrightarrow{OA}$, then

$$\overrightarrow{OC} = j(j\overrightarrow{OA})$$
$$= j^2\overrightarrow{OA}$$
$$= -\overrightarrow{OA} \qquad \text{since} \quad j^2 = -1$$

This is true since it may be seen from the vector diagram that vector $\overrightarrow{OC}$ is equal in magnitude, but opposite in direction, to vector $\overrightarrow{OA}$.

Consider now the effect of the j-operator on the complex number $5 + j3$.

In equation form this is

$$j(5 + j3) = j5 + j(j3)$$
$$= j5 + j^2 3$$
$$= j5 + (-1)3$$
$$= -3 + j5$$

If phasor $z_1 = 5 + j3$ and phasor $z_2 = -3 + j5$, it may be seen from the Argand diagram in Fig. 13.6 that their magnitudes are the same but the effect of the operator j on z_1 has been to alter its direction by $90°$ anti-clockwise to give phasor z_2.

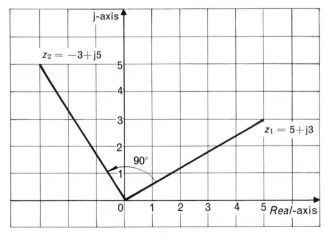

Fig. 13.6

ADDITION OF PHASORS

Consider the addition of the two complex numbers $2+j3$ and $4+j2$.

We have
$$(2+j3)+(4+j2) = 2+j3+4+j2$$
$$= (2+4)+j(3+2)$$
$$= 6+j5$$

On the Argand diagram shown in Fig. 13.7, the complex number $2+j3$ is represented by the phasor $\overrightarrow{OA}$, whilst $4+j2$ is represented by phasor $\overrightarrow{OB}$. The addition of the real parts is performed along the *real*-axis and the addition of the imaginary parts is carried out on the j-axis.

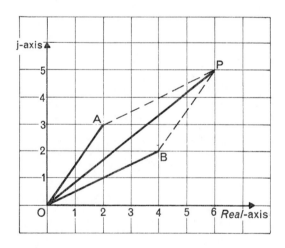

Fig. 13.7

Hence the complex number $6+j5$ is represented by the **phasor** $\overrightarrow{OP}$.

It follows that
$$\overrightarrow{OP} = \overrightarrow{OB} + \overrightarrow{OA}$$

Hence the addition of phasors is similar to vector addition used when dealing with forces or velocities.

SUBTRACTION OF PHASORS

Consider the difference of the two complex numbers, $4 + j5$ and $1 + j4$.

We have
$$(4 + j5) - (1 + j4) = 4 + j5 - 1 - j4$$
$$= (4 - 1) + j(5 - 4)$$
$$= 3 + j$$

On the Argand diagram shown in Fig. 13.8, the complex number $4 + j5$ is represented by the phasor $\overrightarrow{OC}$, whilst $1 + j4$ is represented by the phasor $\overrightarrow{OD}$. The subtraction of the real parts is performed along the *real*-axis, and the subtraction of the imaginary parts is carried out along the j-axis. Now let $(4 + j5) - (1 + j4) = 3 + j$ be represented by the phasor $\overrightarrow{OQ}$.

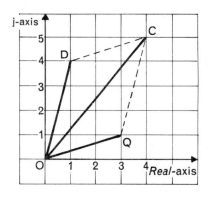

Fig. 13.8

It follows that
$$\overrightarrow{OQ} = \overrightarrow{OC} - \overrightarrow{OD}$$

As for phasor addition, the subtraction of phasors is similar to the subtraction of vectors.

THE POLAR FORM OF A COMPLEX NUMBER

Let z denote the complex number represented by the phasor $\overrightarrow{OP}$ shown in Fig. 13.9. Then from the right-angled triangle PMO we have
$$z = x + jy$$
$$= r \cos \theta + j(r \sin \theta)$$
$$= r(\cos \theta + j \sin \theta)$$

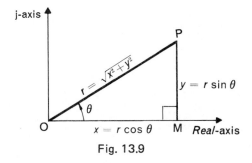

Fig. 13.9

The expression $r(\cos\theta + j\sin\theta)$ is known as the *polar form* of the complex number z. Using conventional notation it may be shown abbreviated as $r\underline{/\theta}$.

r is called the *modulus* of the complex number z and is denoted by $\bmod z$ or $|z|$.

Hence, from the diagram, $|z| = r = \sqrt{x^2 + y^2}$, using the theorem of Pythagoras for right-angled triangle PMO.

It should be noted that the plural of *modulus* is *moduli*.

The angle θ is called the *argument* (or amplitude) of the complex number z, and is denoted by $\arg z$ (or $\operatorname{amp} z$).

Hence $$\arg z = \theta$$

and, from the diagram $$\tan\theta = \frac{y}{x}$$

There are an infinite number of angles whose tangents are the same, and so it is necessary to define which value of θ to state when solving the equation $\tan\theta = \frac{y}{x}$. It is called the principal value of the angle and lies between $+180°$ and $-180°$.

We recommend that, when finding the polar form of a complex number, you should sketch it on an Argand diagram. This will help you to avoid a common error of giving an incorrect value of the angle.

EXAMPLE 13.5

Find the modulus and argument of the complex number $3 + j4$ and express the complex number in polar form.

Let $z = 3 + j4$ which is shown in the Argand diagram in Fig. 13.10.

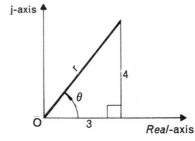

Fig. 13.10

Then $|z| = r = \sqrt{3^2 + 4^2} = 5$

and $\tan \theta = \frac{4}{3} = 1.3333$

$\therefore$ $\theta = 53° 8'$

Hence in polar form

$$z = 5(\cos 53° 8' + j \sin 53° 8')$$

or $z = 5\underline{/53° 8'}$

EXAMPLE 13.6

Show the complex number $z = 5\underline{/-150°}$ on an Argand diagram, and find z in algebraic form.

Now z is represented by phasor $\overrightarrow{OP}$ in Fig. 13.11. It should be noted that since the angle is negative it is measured in a clockwise direction from the *real*-axis datum.

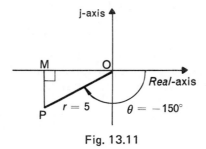

Fig. 13.11

In order to express z in algebraic form we need to find the lengths MO and MP. We use the right-angled triangle PMO in which $P\widehat{O}M = 180° - 150° = 30°$.

Now $\qquad$ MO $=$ PO cos $\widehat{\text{POM}}$ $=$ $5\cos 30°$ $=$ 4.33

and $\qquad$ MP $=$ PO sin POM $=$ $5\sin 30°$ $=$ 2.50

Hence, in algebraic form, the complex number $z = -4.33 - j2.50$

MULTIPLYING NUMBERS IN POLAR FORM

Consider the complex number $z_1 = r_1\underline{/\theta_1} = r_1(\cos\theta_1 + j\sin\theta_1)$

and another complex number $z_2 = r_2\underline{/\theta_1} = r_2(\cos\theta_2 + j\sin\theta_2)$

Then the product of these two complex numbers is

$$
\begin{aligned}
z_1 \times z_2 &= r_1(\cos\theta_1 + j\sin\theta_1) \times r_2(\cos\theta_2 + j\sin\theta_2) \\
&= r_1 r_2(\cos\theta_1 + j\sin\theta_1)(\cos\theta_2 + j\sin\theta_2) \\
&= r_1 r_2\{\cos\theta_1\cos\theta_2 + j\sin\theta_1\cos\theta_2 \\
&\qquad + j\cos\theta_1\sin\theta_2 + j^2\sin\theta_1\sin\theta_2\} \\
&= r_1 r_2\{(\cos\theta_1\cos\theta_2 - \sin\theta_1\sin\theta_2) \\
&\qquad + j(\sin\theta_1\cos\theta_2 + \cos\theta_1\sin\theta_2)\} \\
&= r_1 r_2\{\cos(\theta_1 + \theta_2) + j\sin(\theta_1 + \theta_2)\}† \\
&= r_1 r_2\underline{/\theta_1 + \theta_2)}
\end{aligned}
$$

> Hence to multiply two complex numbers we multiply their moduli and add their arguments.

For example $\quad 6\underline{/17°} \times 3\underline{/35°} = 6 \times 3\underline{/17° + 35°}$

$$
= 18\underline{/52°}
$$

SQUARE ROOT OF A COMPLEX NUMBER

Let $\qquad z = r\underline{/\theta}$

$$
= (\sqrt{r})(\sqrt{r})\underline{\left/\frac{\theta}{2} + \frac{\theta}{2}\right.}
$$

and using the fact that $r_1 r_2\underline{/\theta_1 + \theta_2} = (r_1\underline{/\theta_1}) \times (r_2\underline{/\theta_2})$

then $\qquad z = \left(\sqrt{r}\,\underline{\left/\frac{\theta}{2}\right.}\right) \times \left(\sqrt{r}\,\underline{\left/\frac{\theta}{2}\right.}\right)$

---

†See page 201 for full explanation of the compound angle formulae.

$\therefore$ $$z = \left(\sqrt{r}\;\underline{\left|\frac{\theta}{2}\right.}\right)^2$$

$\therefore$ $$\sqrt{z} = \sqrt{r}\;\underline{\left|\frac{\theta}{2}\right.}$$

Thus

The square root of a complex number is another complex number having a modulus equal to the square root of the given modulus, and amplitude half that of the given argument.

Thus the square root of $36\underline{/24°}$ is $\sqrt{36}\;\underline{\left|\dfrac{24°}{2}\right.}$ or $6\underline{/12°}$.

This is not the only square root but further discussion of these is beyond the scope of our studies at this stage.

If the square root of a complex number in algebraic form is required the number should first be put in polar form and the square root found as above. If needed, the square root may then be expressed in algebraic form.

DIVIDING NUMBERS IN POLAR FORM

It can be shown that the division of two complex numbers, using a method similar to that for finding the product of two complex numbers, is given by

$$\frac{z_1}{z_2} = \frac{r_1\underline{/\theta_1}}{r_2\underline{/\theta_2}} = \frac{r_1}{r_2}\underline{/\theta_1 - \theta_2}$$

Hence if we divide two complex numbers we divide their moduli and subtract their arguments.

For example

$$\frac{5\underline{/33°\,55'}}{3\underline{/-23°\,40'}} = \frac{5}{3}\underline{/(33°\,55') - (-23°\,40')}$$

$$= 1.67\underline{/33°\,55' + 23°\,40'}$$

$$= 1.67\underline{/57°\,35'}$$

EXAMPLE 13.7

A simple circuit which has a resistance R in series with an inductive reactance X_L has an impedance Z is given by the complex number

$$Z = R + jX_L$$

A simple circuit which has a resistance R in series with a capacitive reactance X_C has an impedance Z given by the complex number

$$Z = R - jX_C$$

Using the above relationships, find the resistance and the inductive or capacitive reactance for each of the following impedances:

a) $8 + j12$,

b) $20 - j80$,

c) $40\underline{/25^\circ}$,

d) $100\underline{/-20^\circ}$.

a) Here $Z = 8 + j12$, and since it is of the form $Z = R + jX_L$ we can say that the resistance $R = 8$ and the inductive reactance $X_L = 12$

b) Here $Z = 20 - j80$, and since it is of the form $Z = R - jX_C$ we can say that the resistance $R = 20$ and the capacitive reactance $X_C = 80$

c) The complex number $Z = 40\underline{/25^\circ}$ is shown on the Argand diagram in Fig. 13.12. If we let $Z = x + jy$, then from the diagram

$$x = 40 \cos 25^\circ \quad \text{and} \quad y = 40 \sin 25^\circ$$
$$= 36.3 \qquad\qquad\qquad = 16.9$$

Hence $Z = 36.3 + j16.9$ which is of the form $Z = R + jX_L$ and we can say that resistance $R = 36.3$ and the inductive reactance $X_L = 16.9$

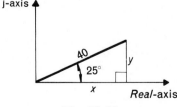

Fig. 13.12

d) The complex number $Z = 100\underline{/-20°}$ is shown on the Argand diagram in Fig. 13.13. If we let $Z = x + jy$, then from the diagram

$$x = 100\cos 20° \quad \text{and} \quad y = 100\sin 20°$$
$$= 94.0 \qquad\qquad = 34.2$$

but we can see from the diagram that the y value is negative hence $Z = 94.0 - j34.2$, which is of the form $Z = R - jX_C$ and we can say that the resistance $R = 94.0$ and the capacitive reactance $X_C = 34.2$

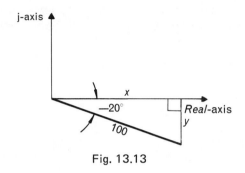

Fig. 13.13

EXAMPLE 13.8

The potential difference across a circuit is given by the complex number $V = 50 + j30$ volts, and the current is given by the complex number $I = 9 + j4$ amperes. Find:

a) the phase difference (i.e. the angle ϕ in Fig. 13.14) between the phasors for V and I,

b) the power, given that power $= |V| \times |I| \times \cos\phi$ watts.

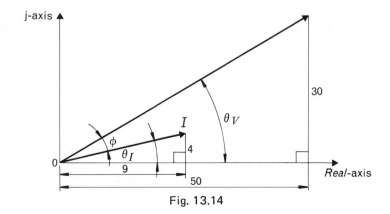

Fig. 13.14

Fig. 13.14 shows a sketch of the Argand diagram showing the phasors for I and V. Phasors in electrical work are usually shown with arrows.

To find $V = 50 + j30$ in polar form

$$|V| = \sqrt{50^2 + 30^2} \quad \text{and} \quad \tan \theta_V = \frac{30}{50}$$

$$= 58.3 \qquad \therefore \qquad \theta_V = 30°\,58'$$

To find $I = 9 + j4$ in polar form

$$|I| = \sqrt{9^2 + 4^2} \quad \text{and} \quad \tan \theta_I = \frac{4}{9}$$

$$= 9.8 \qquad \therefore \qquad \theta_I = 23°\,58'$$

a) The phase difference

$$\phi = \theta_V - \theta_I$$
$$= 30°\,58' - 23°\,58'$$
$$= 7°$$

b) $\qquad$ Power $= |V| \times |I| \times \cos \phi$

$$= 58.3 \times 9.8 \times \cos 7°$$

$$= 567 \text{ watts}$$

Exercise 13.2

1) Show, indicating each one clearly, the following complex numbers on a single Argand diagram: $4 + j3$, $-2 + j$, $3 - j4$, $-3.5 - j2$, $j3$ and $-j4$.

2) Find the moduli and arguments of the complex numbers $3 + j4$ and $4 - j3$.

3) If the complex number $z_1 = -3 + j2$ find $|z_1|$ and $\arg z_1$.

4) If the complex number $z_2 = -4 - j2$ find $|z_2|$ and $\arg z_2$.

5) Express each of the following complex numbers in polar form:

(a) $4 + j3$ $\qquad$ (b) $3 - j4$ $\qquad$ (c) $-3 + j3$ $\qquad$ (d) $-2 - j$

(e) $j4$ $\qquad$ (f) $-j3.5$

6) Convert the following complex numbers, which are given in polar form, into Cartesian form:

(a) $3\underline{/45°}$ (b) $5\underline{/154°}$ (c) $4.6\underline{/-20°}$ (d) $3.2\underline{/-120°}$

7) Simplify the following products of two complex numbers, given in polar form, expressing the answer in polar form:

(a) $8\underline{/30°}\times 7\underline{/40°}$ (b) $2\underline{/-20°}\times 5\underline{/-30°}$

(c) $5\underline{/120°}\times 3\underline{/-30°}$ (d) $7\underline{/-50°}\times 3\underline{/-40°}$

8) Simplify the following divisions of two complex numbers, given in polar form, expressing the answer in polar form:

(a) $\dfrac{8\underline{/20°}}{3\underline{/50°}}$ (b) $\dfrac{10\underline{/-40°}}{5\underline{/20°}}$ (c) $\dfrac{3\underline{/-15°}}{5\underline{/-6°}}$ (d) $\dfrac{1.7\underline{/35°17'}}{0.6\underline{/-9°22'}}$

9) Three complex numbers z_1, z_2 and z_3 are given in polar form by $z_1 = 3\underline{/35°}$, $z_2 = 5\underline{/28°}$ and $z_3 = 2\underline{/-50°}$. Simplify:

(a) $z_1 \times z_2 \times z_3$ giving the answer in polar form.

(b) $\dfrac{z_1 \times z_2}{z_3}$ giving the answer in algebraic form.

10) If the complex number $z = 2 - j3$ express in polar form:

(a) $\dfrac{1}{z}$ (b) z^2

11) Find the square root of:

(a) $8\underline{/38°}$ in polar form.

(b) $2 + j3$ in algebraic form.

12) The admittance Y of a circuit is given by $Y = \dfrac{1}{Z}$.

(a) If $Z = 3 + j5$ find Y in polar form.

(b) If $Z = 17.4\underline{/42°}$ find Y in algebraic form.

13) Using the notation and information given in the data for the worked Example 13.7 of the text (p. 180), find the resistance and the inductive or capacitive reactance for each of the following impedances:

(a) $4.5 + 2.2j$ (b) $23 - 35j$

(c) $29.6\underline{/23°22'}$ (d) $7\underline{/-12°}$

14) The potential difference across a circuit is given by the complex number $V = 40 + j35$ volts and the current is given by the complex number $I = 6 + j3$ amperes. Sketch the appropriate phasors on an Argand diagram and find:

(a) the phase difference (i.e. the angle ϕ) between the phasors for V and I;

(b) the power, given that power $= |V| \times |I| \times \cos \phi$.

14. TRIGONOMETRY WAVEFORMS

On reaching the end of this chapter you should be able to:

1. State the approximations for sin x, cos x and tan x when x is small.
2. Sketch the graphs of

$$\sin A, \sin 2A, 2\sin A \text{ and } \sin \frac{A}{2}$$

and $\quad \cos A, \cos 2A, 2\cos A \text{ and } \cos \frac{A}{2}$

for values of A between 0° and 360°.
3. Sketch the graphs of $\sin^2 A$ and $\cos^2 A$ for values of A between 0° and 360°.
4. Distinguish between angular and time bases.
5. Define and identify amplitude and frequency.
6. Define angular velocity and period.
7. Sketch the graphs of the functions in 2. where A is replaced by ωt.
8. Determine the single wave resulting from a combination of two waves of the same frequency using phasors and a graphical method.
9. Define the term 'phase angle'.
10. Measure the amplitude and phase angle of the resultant wave in 8.
11. Determine graphically the single wave resulting from the combination of two waves (within the limitations of 2.).
12. Show that the resultant of two sine waves of different frequencies gives rise to a non-sinusoidal periodic function.

RADIANS AND DEGREES

We know that one full revolution is equivalent to 360° or 2π radians.

Hence

$$1 \text{ radian} = \left(\frac{360}{2\pi}\right)^{\circ} = \left(\frac{180}{\pi}\right)^{\circ}$$

If tables are used when finding trigonometrical ratios, such as sines and cosines of angles, it may be necessary to convert an angle in radians to an angle in degrees.

For example

$$\sin 0.5 = \sin\left(0.5 \times \frac{180}{\pi}\right)^{\circ} = \sin 28.65° = 0.4795$$

It should be noted that if the units of an angle are omitted it is assumed that it is given in radians (as in the above example).

If a scientific calculator is available it is often possible to set the machine to accept radians by setting a special key. There is then no necessity to convert from radians to degrees.

APPROXIMATIONS FOR TRIGONOMETRICAL FUNCTIONS OF SMALL ANGLES

If the angle θ is small, less than about 0.1 radians or 6°, the following approximations are often used:

$\sin \theta \simeq \theta$	
$\cos \theta \simeq 1 - \dfrac{\theta^2}{2}$	Providing θ is in *radians*
$\tan \theta \simeq \theta$	

For example, when $\theta = 0.1 \, \text{rad} \, (5.73°)$:

Function	Correct to four significant figures	
	True value from calculator	*Approximate value*
$\sin 0.1$	0.0998	0.1000
$\cos 0.1$	0.9950	$1 - \dfrac{(0.1)^2}{2} = 0.9950$
$\tan 0.1$	0.1003	0.1000

AMPLITUDE OR PEAK VALUE

The graphs of $\sin \theta$ and $\cos \theta$ each have a maximum value of $+1$ and a minimum value of -1.

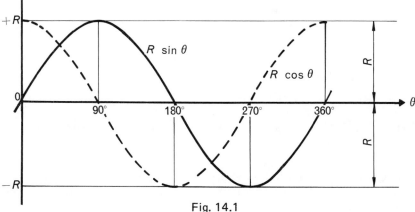

Fig. 14.1

Similarly the graphs of $R \sin \theta$ and $R \cos \theta$ each have a maximum value of $+R$ and a minimum value of $-R$. These graphs are shown in Fig. 14.1 (p. 187).

The value of R is known as the amplitude or peak value.

GRAPHS OF sin θ, sin 2θ, 2 sin θ AND sin $\dfrac{\theta}{2}$

Curves of the above trigonometrical functions are shown plotted in Fig. 14.2. You may find it useful to construct the curves using values obtained from a calculator.

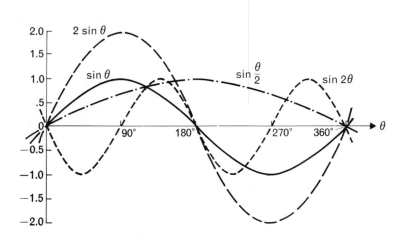

Fig. 14.2

GRAPHS OF cos θ, cos 2θ, 2 cos θ AND cos $\dfrac{\theta}{2}$

Cosine graphs are similar in shape to sine curves. You should plot graphs of the above functions from $0°$ to $360°$ using values obtained from a calculator.

GRAPHS OF sin²θ AND cos²θ

It is sometimes necessary in engineering applications, such as when finding the root mean square value of alternating currents and voltages, to be familiar with the curves $\sin^2\theta$ and $\cos^2\theta$.

Values of the functions are obtained from a calculating machine and their graphs are shown in Fig. 14.3. We should note that the curves are wholly positive, since squares of negative or positive values are always positive.

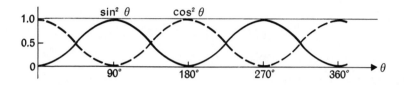

Fig. 14.3

RELATION BETWEEN ANGULAR AND TIME SCALES

In Fig. 14.4 OP represents a radius, of length R, which rotates at a uniform angular velocity ω radians per second about O, the direction of rotation being anticlockwise.

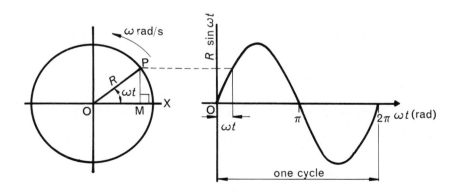

Fig. 14.4

Now

$$\text{Angular velocity} = \frac{\text{Angle turned through}}{\text{Time taken}}$$

∴   Angle turned through = (Angular velocity) × (Time taken)

and hence after a time t seconds

Angle turned through = ωt radians

Also from the right angled triangle OPM:

$$\frac{\text{PM}}{\text{OP}} = \sin \widehat{\text{POM}}$$

∴   PM = OP $\sin \widehat{\text{POM}}$

or   PM = $R \sin \omega t$

If a graph is drawn, as in Fig. 14.4, showing how PM varies with the angle ωt the sine wave representing $R \sin \omega t$ is obtained. It can be seen that the peak value of this sine wave is R (i.e. the magnitude of the rotating radius).

The horizontal scale shows the angle turned through, ωt, and the waveform is said to be plotted on an *angular* or ωt *base.*

CYCLE

A cycle is the portion of the waveform which shows its complete shape without any repetition. It may be seen from Fig. 14.4 that one cycle is completed whilst the radius OP turns through 360° or 2π radians.

PERIOD

This is the time taken for the waveform to complete one cycle.

It will also be the time taken for OP to complete one revolution or 2π radians.

Now we know that

$$\text{Time taken} = \frac{\text{Angle turned through}}{\text{Angular velocity}}$$

Hence

$$\text{The period} = \frac{2\pi}{\omega} \text{ seconds}$$

FREQUENCY

The number of cycles per second is called the frequency. The unit of frequency representing one cycle per second is the hertz (Hz).

Now if   1   cycle is completed in $\dfrac{2\pi}{\omega}$ seconds (a period)

then $1 \div \dfrac{2\pi}{\omega}$ cycles are completed in 1 second

and therefore $\dfrac{\omega}{2\pi}$ cycles are completed in 1 second

Hence

$$\text{Frequency} = \frac{\omega}{2\pi} \text{ Hz}$$

Also since $$\text{Period} = \frac{2\pi}{\omega} \text{ s}$$

$$\text{Frequency} = \frac{1}{\text{Period}}$$

GRAPHS OF sin t, sin $2t$, 2 sin t, AND sin $\frac{1}{2}t$

Now   the waveform $\sin \omega t$ has a period of $\dfrac{2\pi}{\omega}$ seconds

Thus   the waveform $\sin t$ has a period of $\dfrac{2\pi}{1} = 6.28$ seconds

We have seen how a graph may be plotted on an 'angular' or 'ωt' base as in Fig. 14.4. Alternatively the units on the horizontal axis may be those of time (usually seconds), and this is called a 'time' base, as on an oscilloscope.

In order to plot one complete cycle of the waveform it is necessary to take values of t from 0 to 6.28 seconds. We suggest that you should plot the curve of $\sin t$, remembering to set your calculator to the 'radian' mode when finding the volume of $\sin t$. The curve is shown plotted on a time base in Fig. 14.4.

Similarly,

the waveform $\sin 2t$ has a period of $\dfrac{2\pi}{2} = 3.14$ seconds

and the waveform $\sin \frac{1}{2}t$ has a period of $\dfrac{2\pi}{\frac{1}{2}} = 12.56$ seconds

Each of the above waveforms has an amplitude of unity. However the waveform $2\sin t$ has an amplitude of 2, although its period is the same as that of $\sin t$, namely 6.28 seconds.

All these curves are shown plotted in Fig. 14.5. This enables a visual comparison to be made and it may be seen, for example, that the curve of $\sin 2t$ has a frequency twice that of $\sin t$ (since two cycles of $\sin 2t$ are completed during one cycle of $\sin t$).

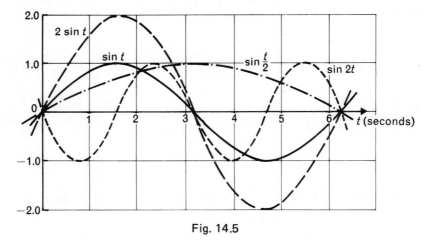

Fig. 14.5

GRAPHS OF $R \cos \omega t$

The waveforms represented by $R \cos \omega t$ are similar to sine waveforms, R being the peak value and $\dfrac{2\pi}{\omega}$ the period. You are left to plot these as instructed in Exercise 14.1 which follows this text.

Exercise 14.1

1) On the same axes, using a time base, plot the waveforms of $\cos t$ and $2\cos t$ for one complete cycle from $t = 0$ to $t = 6.28$ seconds.

2) Using the same axes on which the curves were plotted in Question 1, plot the waveforms of $\cos 2t$ and $\cos \dfrac{t}{2}$.

3) On the same axes, using an angle base from $0°$ to $360°$, sketch the following waveforms:

(a) $5\cos\theta$ (b) $3\sin 2\theta$ (c) $4\cos 3\theta$ (d) $2\sin 3\theta$

PHASE ANGLE

The principal use of sine and cosine waveforms occurs in electrical technology where they represent alternating currents and voltages. In a diagram such as shown in Fig. 14.6 the rotating radii OP and OQ are called phasors.

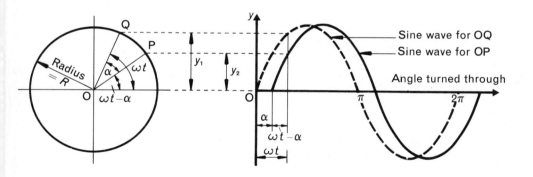

Fig. 14.6

Fig. 14.6 shows two phasors OP and OQ, separated by an angle α, rotating at the same angular speed in an anticlockwise direction. The sine waves produced by OP and OQ are identical curves but they are displaced from each other. The amount of displacement is known as the phase difference and, measured along the horizontal axis, is α. The angle α is called the *phase angle*.

In Fig. 14.6 the phasor OP is said to *lag* behind phasor OQ by the angle α. If the radius of the phasor circle is R then OP = OQ = R and hence

for the phasor OQ, $y_1 = R \sin \omega t$

and for the phasor OP,

$$y_2 = R \sin (\omega t - \alpha)$$

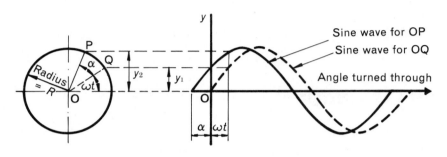

Fig. 14.7

Similarly in Fig. 14.7 the phasor OP leads the phasor OQ by the phase angle, α.

Hence for the phasor OQ,

$$y_1 = R \sin \omega t$$

and for the phasor OP,

$$y_2 = R \sin (\omega t + \alpha)$$

In practice it is usual to draw waveform on an 'angular' or 'ωt' bases, when considering phase angles as in the following example.

EXAMPLE 14.1

Sketch the waveforms of $\sin \omega t$ and $\sin \left(\omega t - \dfrac{\pi}{3}\right)$ on an angular base and identify the phase angle.

The curve $\sin \omega t$ will be plotted between when $\omega t = 0$ and when $\omega t = 2\pi$ radians (i.e. over 1 cycle).

Also $\sin\left(\omega t - \dfrac{\pi}{3}\right)$ will be plotted between values given by

$$\omega t - \frac{\pi}{3} = 0 \quad \text{and when} \quad \omega t - \frac{\pi}{3} = 2\pi$$

i.e. $\qquad \omega t = \dfrac{\pi}{3}$ radians and $\quad \omega t = 2\pi + \dfrac{\pi}{3} = \dfrac{7\pi}{3}$ radians

The graphs are shown in Fig. 14.8.

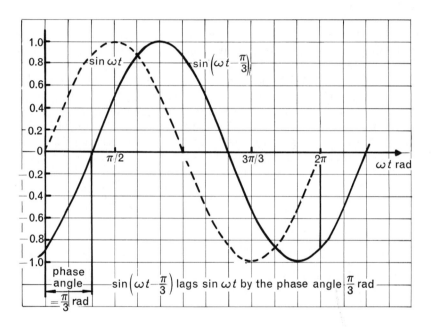

Fig. 14.8

COMBINING SINE WAVES

In the context of the work which follows *combining sine waves* means *adding two sine waves together*. The result of this addition is called the *resultant waveform*. We will examine the result of adding two sine waves of the same frequency and also of adding two sine waves of different frequencies.

Adding two sine waves of the same frequency

The methods used are well illustrated by means of a typical example.

Consider two sinusoidal electric currents represented by the equations $i_1 = 10 \sin \theta°$ and $i_2 = 12 \sin (\theta + 30)°$. If the resultant waveform is denoted by i_r, then

$$i_r = i_1 + i_2 = 10 \sin \theta° + 12 \sin (\theta + 30)°$$

The addition may be achieved by either phasor addition or addition of the sine waves.

PHASOR ADDITION   The addition of phasors is similar to vector addition used when dealing with forces or velocities. The curves of i_1 and i_2 are shown in Fig. 14.9 together with their associated phasors OC' and OD'. The angle $C'OD'$ of $30°$ between these phasors represents the phase angle by which i_2 leads i_1. In order to add phasors OC' and OD' we construct the parallelogram $OC'B'D'$. Then diagonal OB' gives the resultant phasor. The amplitude or peak value of i_r is given by the length of OB' which is 21.3, found by measurement. The phase angle between i_r and i_1 is given by the angle $C'OB'$. Thus i_r leads i_1 by $16°$, found by measurement.

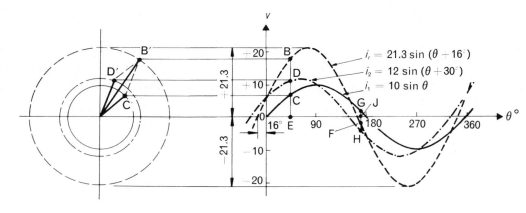

Fig. 14.9

ADDITION OF THE SINE WAVES   In order to plot the waveform for i_r we need values of the expression $10 \sin \theta° + 12 \sin (\theta + 30)°$ for suitable values of θ from $0°$ to $360°$. A suitable calculator sequence using, for example, $\theta = 25°$ is as follows, remembering to set first the angle mode on the machine to 'degrees'.

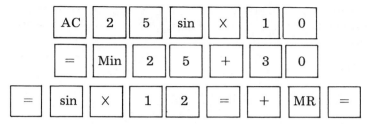

giving an answer 14.1 correct to three significant figures.

The graphs of i_1, i_2 and i_r are shown in Fig. 14.9 where it will be seen that i_r is a sine wave with a peak value of 21.3 and a phase angle of $16°$. It has the equation $i_r = 21.3 \sin (\theta + 16°)$.

The peak value of 21.3 and the phase angle of $16°$ are only approximate as their accuracy depends on reading values from the scales of the graphs. In this case it is possible to calculate these values by a theoretical method and obtain more accurate results. The method is as shown on p. 209 and gives the more accurate answers of 21.254 and $16°24'$ which shows the answers we obtained are as good as could be expected from a graphical method.

It is possible to obtain the i_r curve from the graphs of i_1 and i_2 by graphical addition. This is, in fact, similar to adding the values of i_1 and i_2 for various values of the angle θ.

First plot the graphs of i_1 and i_2 and then the points on the i_r curve may be plotted by adding the ordinates (i.e. the vertical lengths) of the corresponding points on the i_1 and i_2 curves.

For example to plot point B on the i_r curve in Fig. 14.9 we can measure the lengths CE and DE and then add their values, since BE = CE + DE.

We must take care to allow for the ordinates being positive or negative. For example to find point F on the i_r curve we must use FJ = GJ − HJ.

Adding two sine waves of different frequencies

Since we desire to see the shape of the resultant curve, we will add the given sine waves. Let us see the result of adding together the sine curves $\sin \theta$ and $\sin 2\theta$. Note that $\sin 2\theta$ has a frequency twice that of $\sin \theta$.

In order to plot the resultant curve we need values of the expression

$(\sin \theta + \sin 2\theta)$. As usual we will use values of θ from $0°$ to $360°$. A suitable calculator sequence using, for example, $\theta = 60°$ is as follows:

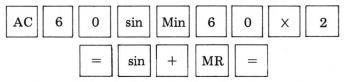

giving an answer 1.73 correct to three significant figures.

The curves of $\sin \theta$, $\sin 2\theta$, and $(\sin \theta + \sin 2\theta)$ are shown plotted in Fig. 14.10.

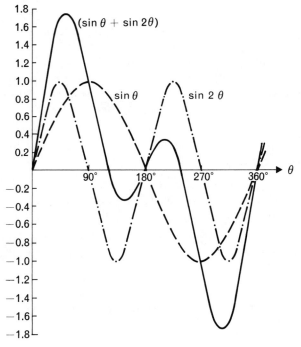

Fig. 14.10

We can see that the graph of $(\sin \theta + \sin 2\theta)$ is non-sinusoidal (i.e. *not* the shape of a sine curve). However, the curve is a waveform (or a periodic function) since it will repeat indefinitely.

In general the addition of any two sine waves of different frequencies will result in a non-sinusoidal waveform. You may check this by working through the examples in Question 10 of Exercise 14.2 which follows.

Exercise 14.2

1) Plot the graphs of $\sin \theta$ and $\sin (\theta + 0.9)$ on the same axes on an angular base using units in radians. Indicate the phase angle between the waveforms and explain whether it is an angle of lead or lag.

2) Plot the graph of $\sin \left(\omega t + \dfrac{\pi}{6}\right)$ and $\sin \omega t$ on the same axes on an angular base showing a cycle of each waveform. Identify the phase angle between the curves.

3) Plot the graphs of $\sin \left(\omega t + \dfrac{\pi}{3}\right)$ and $\sin \left(\omega t - \dfrac{\pi}{4}\right)$ on the same axes on an angular base showing a cycle of each waveform. Identify the phase angle between the curves.

4) Write down the equation of the waveform which:

(a) leads $\sin \omega t$ by $\dfrac{\pi}{2}$ radians.

(b) lags $\sin \omega t$ by π radians.

(c) leads $\sin \left(\omega t - \dfrac{\pi}{3}\right)$ by $\dfrac{\pi}{3}$ radians.

(d) lags $\sin \left(\omega t + \dfrac{\pi}{6}\right)$ by $\dfrac{\pi}{3}$ radians.

5) Plot the waves of $\sin \theta$ and $\cos \theta$ on the same axes on an angular base showing a cycle of each waveform. Identify the phase angle between the curves.

6) Find the resultant waveform of the curves $3 \sin \theta$ and $2 \cos \theta$ by graphical addition. What is the amplitude of the resultant waveform and the phase angle relative to $3 \sin \theta$?

7) Find the resultant voltage v_R of the two voltages represented by the equations $v_1 = 3 \sin \theta$ and $v_2 = 5 \sin (\theta - 30°)$ by plotting the three graphs.

8) Plot the graphs of $i_1 = 5 \sin \theta$ and $i_2 = 2 \sin (\theta + 45°)$ and hence find the resultant of i_1 and i_2 by graphical addition. State the equation of the resultant current i_R.

9) The voltages v_1 and v_2 are represented by the equations $v_1 = 30 \sin (\theta + 60°)$ and $v_2 = 50 \sin (\theta - 45°)$. Plot the curves of v_1 and v_2 and the resultant voltage v_R and find the equation representing v_R.

10) Find the resultant curve in each of the following and verify that the resulta are non-sinusoidal periodic waveforms:

(a) $\sin\theta + \sin\dfrac{\theta}{2}$

(b) $3\sin\theta + 2\sin 2\theta$

(c) $\sin\dfrac{\theta}{2} + 2\sin 2\theta$

(d) $\sin(\theta + 30°) + \sin 2\theta$

15. COMPOUND-ANGLE FORMULAE

On reaching the end of this chapter you should be able to:

1. *Use the formulae for sin (A ± B), cos (A ± B), tan (A + B).*
2. *Derive the double-angle formulae for sin 2A, cos 2A, tan 2A.*
3. *Express R sin (ωt ± α) in the form*

   $$a\sin \omega t \pm b\cos \omega t$$

   using 1. and vice-versa.
4. *Deduce the relationship between a, b, R and α.*

INTRODUCTION

It can be shown that

$$\sin (A + B) = \sin A \cos B + \cos A \sin B \qquad [1]$$

$$\sin (A - B) = \sin A \cos B - \cos A \sin B \qquad [2]$$

$$\cos (A + B) = \cos A \cos B - \sin A \sin B \qquad [3]$$

$$\cos (A - B) = \cos A \cos B + \sin A \sin B \qquad [4]$$

On inspecting these formulae you may feel that an error has been made in the 'signs' on the right-hand sides of the equations for $\cos (A \pm B)$. They are correct, however, and a special note should be made of them.

Since the above formulae involve the two angles A and B they are called *compound-angle* formulae.

The following examples show some uses of the above formulae.

EXAMPLE 15.1

Simplify:

a) $\sin (\theta + 90°)$,

b) $\cos (\theta - 270°)$.

a) Using $\sin (A + B) = \sin A \cos B + \cos A \sin B$

and substituting θ for A and $90°$ for B

we have $\quad \sin(\theta + 90°) = \sin\theta \cos 90° + \cos\theta \sin 90°$

$$= (\sin\theta)0 + (\cos\theta)1$$

$$= \cos\theta$$

b) Using

$$\cos(A - B) = \cos A \cos B + \sin A \sin B$$

and substituting θ for A and $270°$ for B

we have

$$\cos(\theta - 270°) = \cos\theta \cos 270° + \sin\theta \sin 270°$$

$$= (\cos\theta)0 + (\sin\theta)(-1)$$

$$= -\sin\theta$$

tan $(A + B)$ FORMULA

Now if we divide equation [1] by equation [3] we have

$$\frac{\sin(A+B)}{\cos(A+B)} = \frac{\sin A \cos B + \cos A \sin B}{\cos A \cos B - \sin A \sin B}$$

and if we divide both numerator and denominator of the right-hand side by $(\cos A \cos B)$

Then $\qquad \tan(A+B) = \dfrac{\dfrac{\sin A \cos B}{\cos A \cos B} + \dfrac{\cos A \sin B}{\cos A \cos B}}{\dfrac{\cos A \cos B}{\cos A \cos B} - \dfrac{\sin A \sin B}{\cos A \cos B}}$

$\therefore \qquad \boxed{\tan(A+B) = \dfrac{\tan A + \tan B}{1 - \tan A \tan B}} \qquad\qquad [5]$

EXAMPLE 15.2

By using the sines, cosines and tangents of $30°$ and $45°$ established from suitable right-angled triangles find:

a) $\sin 75°$,

b) $\cos 15°$,

c) $\tan 75°$.

To find the trigonometrical ratios of $30°$ and $45°$ we use suitable right-angled triangles shown in Figs. 15.1 and 15.2.

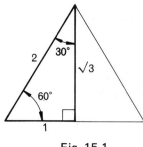

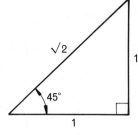

Fig. 15.1   Fig. 15.2

From Fig. 15.1 we have

$$\sin 30° = \frac{1}{2}, \quad \cos 30° = \frac{\sqrt{3}}{2} \quad \text{and} \quad \tan 30° = \frac{1}{\sqrt{3}}$$

and from Fig. 15.2 we have

$$\sin 45° = \frac{1}{\sqrt{2}}, \quad \cos 45° = \frac{1}{\sqrt{2}} \quad \text{and} \quad \tan 45° = 1$$

a) Now

$$\sin(45° + 30°) = \sin 45° \cos 30° + \cos 45° \sin 30°$$

$$\therefore \quad \sin 75° = \left(\frac{1}{\sqrt{2}}\right)\left(\frac{\sqrt{3}}{2}\right) + \left(\frac{1}{\sqrt{2}}\right)\left(\frac{1}{2}\right)$$

$$= \frac{\sqrt{3} + 1}{2\sqrt{2}}$$

$$= 0.966$$

b) Now $\cos(A - B) = \cos A \cos B + \sin A \sin B$

and if we substitute $45°$ for A and $30°$ for B

then $\cos(45° - 30°) = \cos 45° \cos 30° + \sin 45° \sin 30°$

$$\therefore \quad \cos 15° = \left(\frac{1}{\sqrt{2}}\right)\left(\frac{\sqrt{3}}{2}\right) + \left(\frac{1}{\sqrt{2}}\right)\left(\frac{1}{2}\right)$$

$$= \frac{\sqrt{3} + 1}{2\sqrt{2}}$$

$$= 0.966$$

This confirms the fact that

$$\sin 75° = \cos 15°$$

c) Now $\tan(A+B) = \dfrac{\tan A + \tan B}{1 - \tan A \tan B}$

and if we substitute $45°$ for A, and $30°$ for B

then $\tan(45° + 30°) = \dfrac{\tan 45° + \tan 30°}{1 - \tan 45° \tan 30°}$

$\therefore \qquad \tan 75° = \dfrac{1 + 1/\sqrt{3}}{1 - 1(1/\sqrt{3})}$

$\qquad\qquad\qquad = 3.732$

EXAMPLE 15.3

Find the angle between $0°$ and $90°$ which satisfies the equation

$$\sin x = 2\sin(x - 20°)$$

We have $\quad \sin x = 2\sin(x - 20°)$

$\qquad\qquad\quad = 2\{\sin x \cos 20° - \cos x \sin 20°\}$

$\qquad\qquad\quad = 2\{(\sin x)0.9397 - (\cos x)0.3420\}$

$\qquad\qquad\quad = 1.8794(\sin x) - 0.6840(\cos x)$

$\therefore \quad 0.6840\cos x = 1.8794\sin x - \sin x$

$\therefore \quad 0.6840\cos x = 0.8794\sin x$

$\therefore \qquad \dfrac{\sin x}{\cos x} = \dfrac{0.6840}{0.8794}$

$\therefore \qquad \tan x = 0.7778$

$\therefore \qquad x = 37°53'$

DOUBLE-ANGLE FORMULAE

The term *double angle* refers to angle $2A$. A double-angle formula expresses a trigonometrical ratio of a double angle in terms of trigonometrical ratio(s) of the single angle.

sin 2A formula

If we put $B = A$ into equation [1] then we have

$$\sin(A + A) = \sin A \cos A + \cos A \sin A$$

or $\qquad \boxed{\sin 2A = 2\sin A \cos A} \qquad\qquad\qquad$ [6]

cos 2*A* formulae

If we put $B = A$ into equation [3], then we have

$$\cos(A + A) = \cos A \cos A - \sin A \sin A$$

or

$$\boxed{\cos 2A = \cos^2 A - \sin^2 A} \qquad [7]$$

For any angle we know that $\sin^2 A + \cos^2 A = 1$

from which either $\qquad\qquad\qquad\qquad \sin^2 A = 1 - \cos^2 A$

or $\qquad\qquad\qquad\qquad\qquad\qquad \cos^2 A = 1 - \sin^2 A$

Now substituting $\sin^2 A = 1 - \cos^2 A$ into equation [7], we have

$$\cos 2A = \cos^2 A - (1 - \cos^2 A)$$

or

$$\boxed{\cos 2A = 2\cos^2 A - 1} \qquad [8]$$

Also substituting $\cos^2 A = 1 - \sin^2 A$ into equation [7], we have

$$\cos 2A = (1 - \sin^2 A) - \sin^2 A$$

or

$$\boxed{\cos 2A = 1 - 2\sin^2 A} \qquad [9]$$

tan 2*A* formula

If we put $B = A$ into equation [5], then we have

$$\tan(A + A) = \frac{\tan A + \tan A}{1 - \tan A \tan B}$$

or

$$\boxed{\tan 2A = \frac{2\tan A}{1 - \tan^2 A}} \qquad [10]$$

EXAMPLE 15.4

If $\cos \phi = \frac{24}{25}$ find, without calculating the value of the angle ϕ, the values of:

a) $\sin 2\phi$ b) $\cos 2\phi$ c) $\tan 2\phi$

Using the right-angled triangle ABC shown in Fig. 15.3 and the theorem of Pythagoras

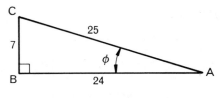

Fig. 15.3

we have $CB^2 = AC^2 - BA^2 = 25^2 - 24^2 = 49$

Thus $CB = 7$

Hence $\sin \phi = \frac{7}{25}$, $\cos \phi = \frac{24}{25}$ and $\tan \phi = \frac{7}{24}$

a) Now $\sin 2\phi = 2 \sin \phi \cos \phi = 2(\frac{7}{25})(\frac{24}{25}) = 0.538$

b) Now $\cos 2\phi = 2 \cos^2 \phi - 1 = 2(\frac{24}{25})^2 - 1 = 0.843$

c) Now $\tan 2\phi = \dfrac{2 \tan \phi}{1 - \tan^2 \phi} = \dfrac{2(7/24)}{1 - (7/24)^2} = 0.638$

Exercise 15.1

1) Simplify:
(a) $\sin (x + 180°)$ (b) $\cos (180° - x)$
(c) $\sin (90° - x)$ (d) $\cos (270° + x)$
(e) $\tan (x + 180°)$

2) Find the value of $\sin 15°$ using the fact $\sin 15° = \sin (45° - 30°)$.

3) Show that $\sin (\phi + 60°) + \sin (\phi - 60°) = \sin \phi$,

4) If $\tan \theta = \frac{3}{4}$ and $\tan \phi = \frac{5}{12}$ sketch suitable right-angled triangles from which the other trigonometrical ratios of θ and ϕ may be found. Hence find the values of:
(a) $\sin (\theta - \phi)$ (b) $\cos (\theta + \phi)$ (c) $\tan (\theta + \phi)$
(d) $\sin 2\theta$ (e) $\cos 2\phi$ (f) $\tan 2\theta$

5) Show that $\sqrt{2} \sin \left(\theta - \dfrac{\pi}{4} \right) = \sin \theta - \cos \theta$.

6) Find the angle between $0°$ and $90°$ which satisfies the equation $\cos \theta = 3 \cos (\theta + 30°)$.

7) Illustrate the facts that $\cos \alpha = \frac{3}{5}$ and $\cos \beta = \frac{4}{5}$ using a right-angled triangle. Without calculating the values of the angles α and β

(a) show that $\alpha + \beta = 90°$,

(b) find the values of $\sin 2\alpha$, $\cos 2\beta$, and $\tan 2\alpha$.

8) Find the angle between $0°$ and $90°$ which satisfies the equation $2 \cos (\theta + 60°) = \sin (\theta + 30°)$.

9) Find the angle between $0°$ and $90°$ which satisfies the equation $(1 + \tan A) \tan 2A = 1$.

10) If $\sin 33° 24' = 0.5505$ find, by using the compound angle formulae, the values of:

(a) $\cos 236° 36'$ (b) $\sin 326° 36'$

THE FORM $R \sin (\theta \pm \alpha)$

Consider $3 \sin (\theta + 40°)$.

$$\text{Now} \qquad \sin (A + B) = \sin A \cos B + \cos A \sin B$$

$$\therefore \qquad 3 \sin (\theta + 40°) = 3[\sin \theta \cos 40° + \cos \theta \sin 40°]$$

$$= 3[(\sin \theta)0.7660 + (\cos \theta)0.6428]$$

$$= 2.298 \sin \theta + 1.928 \cos \theta$$

Now using the same method, if R and α are constants

$$\text{then} \qquad R \sin (\theta \pm \alpha) = R[\sin \theta \cos \alpha \pm \cos \theta \sin \alpha]$$

$$= (R \cos \alpha) \sin \theta \pm (R \sin \alpha) \cos \theta$$

$$= a \sin \theta \pm b \cos \theta$$

where	$a = R \cos \alpha$	[1]
and	$b = R \sin \alpha$	[2]

$\therefore$ squaring and adding equations [1] and [2] gives

$$(R \cos \alpha)^2 + (R \sin \alpha)^2 = a^2 + b^2$$

$$\therefore \qquad R^2 \cos^2\alpha + R^2 \sin^2\alpha = a^2 + b^2$$

$$\therefore \qquad R^2[\cos^2\alpha + \sin^2\alpha] = a^2 + b^2$$

and since $\cos^2\alpha + \sin^2\alpha = 1$

then $$R^2 = a^2 + b^2$$

Also dividing equation [2] by equation [1] gives

$$\frac{R \sin \alpha}{R \cos \alpha} = \frac{b}{a}$$

∴ $$\tan \alpha = \frac{b}{a}$$

Hence

$$R \sin (\theta \pm \alpha) = a \sin \theta \pm b \cos \theta$$

where

$$R^2 = a^2 + b^2 \quad \text{and} \quad \tan \alpha = \frac{b}{a}$$

Problems using the above relationship usually occur in the reverse order to the sequence above.

EXAMPLE 15.5

Express $4 \sin \theta - 3 \cos \theta$ in the form $R \sin (\theta - \alpha)$.

Comparing the given expression with $a \sin \theta - b \cos \theta$ we have $a = 4$ and $b = 3$,

and since $R^2 = a^2 + b^2$ and also $\tan \alpha = \dfrac{b}{a}$

then $R^2 = 4^2 + 3^2$ $\tan \alpha = \dfrac{3}{4}$

∴ $R = \sqrt{16 + 9}$ $= 0.75$

∴ $R = 5$ ∴ $\alpha = 36° 52'$

Hence $4 \sin \theta - 3 \cos \theta = 5 \sin (\theta - 36° 52')$

EXAMPLE 15.6

Express $6 \sin \omega t - 8 \cos \omega t$ in the form $R \sin (\theta - \alpha)$. Hence find the maximum value of $6 \sin \omega t - 8 \cos \omega t$ and the value of ωt at which it occurs.

This example is similar to Example 15.5 except that the angles ωt and α are in radians.

Comparing the given expression with $a \sin \theta + b \cos \theta$ we have $a = 6$ and $b = 8$.

Then $\qquad R = \sqrt{6^2 + 8^2} = 10 \quad$ and $\quad \tan \alpha = \dfrac{8}{6} = 1.333$

$$\alpha = 0.927 \text{ rad}$$

Hence

$$6 \sin \omega t - 8 \cos \omega t = 10 \sin (\omega t - 0.927)$$

Now the maximum value of the sine of an angle is unity, and occurs when the angle is $90°$ or $\dfrac{\pi}{2}$ radians.

Thus $6 \sin \omega t - 8 \cos \omega t$ will be a maximum

when $\quad \sin (\omega t - 0.927) = 1$

i.e. when $\qquad \omega t - 0.927 = \dfrac{\pi}{2} = 1.571$

or $\qquad\qquad\qquad \omega t = 1.571 + 0.927 = 2.498$

Therefore, the maximum value of $6 \sin \omega t - 8 \cos \omega t$ is 10 and it occurs when $\omega t = 2.498$ rad.

EXAMPLE 15.7

Express $10 \sin \theta + 12 \sin (\theta + 30)°$ in the form $R \sin (\theta + \alpha)°$.

Using the expression

$$\sin (A + B) = \sin A \cos B + \cos A \sin B$$

we have

$$\begin{aligned} 10 \sin \theta + 12 \sin (\theta + 30)° &= 10 \sin \theta + 12 \,(\sin \theta \cos 30° \\ &\qquad + \cos \theta \sin 30°) \\ &= 10 \sin \theta + 12 \cos 30° \sin \theta \\ &\qquad + 12 \sin 30° \cos \theta \\ &= 10 \sin \theta + 10.4 \sin \theta + 6 \cos \theta \\ &= 20.4 \sin \theta + 6 \cos \theta \end{aligned}$$

Now comparing this expression with $a \sin \theta + b \cos \theta$ we have

$$R^2 = 20.4^2 + 6^2 \quad \text{and} \quad \tan \alpha = \frac{6}{20.4}$$

$$\therefore \qquad R = 21.3 \qquad\qquad \alpha = 16.3°$$

Thus $10 \sin \theta + 12 \sin (\theta + 30)° = 21.3 \sin (\theta + 16.3)°$

This result confirms the answer obtained graphically on p. 197.

Exercise 15.2

1) Express $3 \sin \theta + 2 \cos \theta$ in the form $R \sin (\theta + \alpha)$.

2) Express $7 \cos \omega t + \sin \omega t$ in the form $R \sin (\omega t + \alpha)$.

3) Rewrite $5 \sin \omega t - 7 \cos \omega t$ in the form $R \sin (\omega t - \alpha)$.

4) Using the result obtained in Question 3 find the maximum value of $5 \sin \omega t - 7 \cos \omega t$ and the value of ωt at which this occurs.

5) An electric current is given by $i = 200 \sin 300t + 100 \cos 300t$. Express this as a single trigonometrical function and find its maximum value.

6) Express $3 \sin \theta + 5 \sin (\theta - 30°)$ in the form $R \sin (\theta + \alpha)$.

7) Express $5 \sin \theta + 2 \sin (\theta + 45°)$ in the form $R \sin (\theta + \alpha)$.

8) Express $30 \sin (\theta + 60°) + 50 \sin (\theta - 45°)$ in the form

$$R \sin (\theta + \alpha)$$

SUMMARY

This may prove useful for reference.

DIFFERENTIAL COEFFICIENTS OF THE MORE COMMON FUNCTIONS

y	$\dfrac{dy}{dx}$
ax^n	anx^{n-1}
$\sin ax$	$a \cos ax$
$\cos ax$	$-a \sin ax$
$\tan x$	$\sec^2 x$
$\log_e x$	$\dfrac{1}{x}$
e^{ax}	ae^{ax}

INTEGRALS OF THE MORE COMMON FUNCTIONS

y	$y\,dx$
ax^n	$\dfrac{a}{n+1}x^{n+1}$
$\sin ax$	$-\dfrac{1}{a}\cos ax$
$\cos ax$	$\dfrac{1}{a}\sin ax$
$\sec^2 x$	$\tan x$
$\dfrac{1}{x}$	$\log_e x$
e^{ax}	$\dfrac{1}{a}e^{ax}$

TWO TESTS FOR MAXIMUM OR MINIMUM

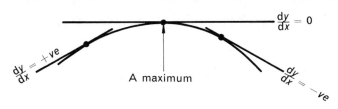

A maximum

211

a) Test for sign of $\dfrac{dy}{dx}$.

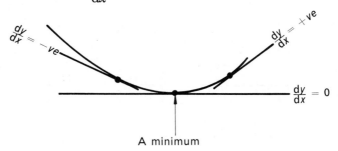

A minimum

b) Test for sign of $\dfrac{d^2y}{dx^2}$.

For a maximum $\dfrac{d^2y}{dx^2} = -ve$

and for a minimum $\dfrac{d^2y}{dx^2} = +ve$

DIFFERENTIATION BY SUBSTITUTION

Use the formula $\dfrac{dy}{dx} = \dfrac{dy}{du} \times \dfrac{du}{dx}$

DIFFERENTIATION OF A PRODUCT

If $y = u \times v$ then $\dfrac{dy}{dx} = v\dfrac{du}{dx} + u\dfrac{dv}{dx}$

DIFFERENTIATION OF A QUOTIENT

If $y = \dfrac{u}{v}$, then $\dfrac{dy}{dx} = \dfrac{v\dfrac{du}{dx} - u\dfrac{dv}{dx}}{v^2}$

MEAN AND ROOT MEAN SQUARE VALUES

$$\text{Mean (or average) value} = \frac{\text{Area under the curve}}{\text{Length of base}}$$

$$\text{Root mean square value} = \sqrt{\frac{\text{Area under the } y^2 \text{ curve}}{\text{Length of base}}}$$

TRAPEZOIDAL RULE

Area under curve = (Strip width)
$\times \frac{1}{2}$(Sum of first and last ordinates)
+ (The sum of remaining ordinates)]

MID-ORDINATE RULE

Area under curve = (Strip width) × (Sum of mid-ordinates)

SIMPSON'S RULE

Area under curve = $\frac{1}{3}$(Strip width)
× [(Sum of first and last ordinates)
+ 4(Sum of even ordinates)
+ 2(Sum of remaining odd ordinates)]

DIFFERENTIAL EQUATION

The differential equation

$$\frac{dy}{dx} = ky$$

has a general solution of the form

$$y = Ae^{kx}$$

LOGARITHMS

Natural logarithms are given as $\log_e$ or ln.

Common logarithms are given as $\log_{10}$ or lg.

$$\log xy = \log x + \log y$$

$$\log \frac{x}{y} = \log x - \log y$$

$$\log x^n = n \log x$$

$$\log_e N = 2.3026 \log_{10} N$$

THE BINOMIAL SERIES

$$(1+x)^n = 1 + nx + \frac{n(n-1)}{2!}x^2 + \frac{n(n-1)(n-2)}{3!} + \ldots$$

THE EXPONENTIAL SERIES

$$e^x = 1 + x + \frac{x^2}{2!} + \frac{x^3}{3!} + \frac{x^4}{4!} + \ldots$$

MATRICES

Addition and subtraction
$$\begin{pmatrix} a & b \\ c & d \end{pmatrix} \pm \begin{pmatrix} p & q \\ r & s \end{pmatrix} = \begin{pmatrix} a \pm p \\ c \pm r \end{pmatrix} \begin{pmatrix} b \pm q \\ r \pm s \end{pmatrix}$$

Zero (or null) matrix
$$\begin{pmatrix} 0 & 0 \\ 0 & 0 \end{pmatrix}$$

Identity (or unit) matrix
$$I = \begin{pmatrix} 1 & 0 \\ 0 & 1 \end{pmatrix}$$

Transpose of matrix. If $A = \begin{pmatrix} a & b & c \\ c & d & e \end{pmatrix}$, then $A' = \begin{pmatrix} a & d \\ b & d \\ c & e \end{pmatrix}$

Multiplication
$$\begin{pmatrix} a & b \\ c & d \end{pmatrix} \times \begin{pmatrix} p & q \\ r & s \end{pmatrix} = \begin{pmatrix} ae+bg & af+bh \\ ce+dg & cf+dh \end{pmatrix}$$

Determinant of matrix. If $A = \begin{pmatrix} a & b \\ c & d \end{pmatrix}$,

then
$$|A| = \begin{vmatrix} a & b \\ c & d \end{vmatrix} = ad - bc$$

Inverse A^{-1} of matrix A is such that $AA^{-1} = \begin{pmatrix} 1 & 0 \\ 0 & 1 \end{pmatrix} = I$

Also if $A = \begin{pmatrix} a & b \\ c & d \end{pmatrix}$, then $A^{-1} = \dfrac{1}{|A|} \begin{pmatrix} d & -b \\ -c & a \end{pmatrix} = \dfrac{\begin{pmatrix} d & -b \\ -c & a \end{pmatrix}}{\begin{vmatrix} a & b \\ c & d \end{vmatrix}}$

COMPLEX NUMBERS

$$j = \sqrt{-1} \quad \text{or} \quad j^2 = -1$$

The algebraic form of a complex number is $x + jy$.

The polar form of a complex number is $r(\cos\theta + j\sin\theta)$ or $r\underline{/\theta}$.

The conjugate of $(x + jy)$ is $(x - jy)$.

The conjugate of $(x - jy)$ is $(x + jy)$.

If a complex number is multiplied by its conjugate, the result is a real number.

The effect of j as an operator on a phasor representing a complex number is to change its direction by $90°$ anti-clockwise, without any alteration in its magnitude.

If a complex number z in its algebraic form is $z = x + jy$.

and in its polar form is

$$z = r(\cos\theta + j\sin\theta) \quad \text{or} \quad r\underline{/\theta}$$

then $\qquad \mod z = |z| = r = \sqrt{x^2 + y^2}$

and $\qquad \arg z = \theta \quad \text{where} \quad \tan\theta = \dfrac{y}{x}$

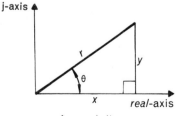

Argand diagram

To multiply two complex numbers in polar form, we multiply their moduli and add their arguments. Thus

$$(r_1 \, \underline{/\theta_1}) \times (r_2 \, \underline{/\theta_2}) \; = \; (r_1 \times r_2) \underline{/\theta_1 + \theta_2}$$

To divide two complex numbers in polar form, we divide their moduli and subtract their arguments. Thus

$$\frac{r_1 \, \underline{/\theta_1}}{r_2 \, \underline{/\theta_2}} \; = \; \left(\frac{r_1}{r_2}\right) \underline{/\theta_1 - \theta_2}$$

TRIGONOMETRY

$$1 \text{ radian} \; = \; \left(\frac{360}{2\pi}\right)^{\circ}$$

For small angles:

$\sin \theta \simeq \theta$	
$\cos \theta \simeq 1 - \dfrac{\theta^2}{1}$	Provided θ is in radians
$\tan \theta \simeq \theta$	

$$\sin (A \pm B) \; = \; \sin A \, \cos B \pm \cos A \, \sin B$$

$$\cos (A \pm B) \; = \; \cos A \, \cos B \mp \sin A \, \sin B$$

$$\tan (A + B) \; = \; \frac{\tan A + \tan B}{1 - \tan A \, \tan B}$$

$$\sin 2A \; = \; 2 \sin A \, \cos A$$

$$\cos 2A \; = \; \cos^2 A - \sin^2 A$$

$$= \; 2 \cos^2 A - 1$$

$$= \; 1 - 2 \sin^2 A$$

$$\tan 2A \; = \; \frac{2 \tan A}{1 - \tan^2 A}$$

$$R \sin (\theta \pm \alpha) \; = \; a \sin \theta \pm b \cos \theta$$

where $\qquad R^2 \; = \; a^2 + b^2 \quad \text{and} \quad \tan \alpha \; = \; \dfrac{b}{a}$

ANSWERS

ANSWERS TO CHAPTER 1

Exercise 1.1

1) $15x^2 + 14x - 1$
2) $3.5t^{-0.5} - 1.8t^{-0.7}$
3) 1.25 4) 1 5) $5, -3$

Exercise 1.2

1) $6(3x + 1)$ 2) $-15(2 - 5x)^2$
3) $-2(1 - 4x)^{-1/2}$
4) $-7.5(2 - 5x)^{1/2}$
5) $-4(4x^2 + 3)^{-2}$
6) $3\cos(3x + 4)$
7) $5\sin(2 - 5x)$
8) $8\sin 4x \cos 4x$
9) $21(\sin 7x)/\cos^4 7x$
10) $2\cos\left(2x + \dfrac{\pi}{2}\right)$
11) $-3(\sin x)\cos^2 x$
12) $-\dfrac{\cos x}{\sin^2 x}$
13) $\dfrac{1}{x}$ 14) $-\dfrac{9}{x}$
15) $\dfrac{1}{2(2x - 7)}$ 16) $-\dfrac{1}{e^x}$
17) $6e^{(3x+4)}$ 18) $8e^{(8x-2)}$
19) $\frac{2}{3}(1 - 2t)^{-4/3}$
20) $\frac{3}{4}\cos(\frac{3}{4}\theta - \pi)$
21) $\{-\sin(\pi - \phi)\}/\cos^2(\pi - \phi)$
22) $-\dfrac{1}{2x}$ 23) $Bke^{(kt-b)}$
24) $-\dfrac{1}{3}e^{(1-x)/3}$

Exercise 1.3

1) (a) $\sin x + x\cos x$
   (b) $e^x(\tan x + \sec^2 x)$
   (c) $1 + \log_e x$

2) $\cos^2 t - \sin^2 t$
3) $2(\tan\theta)\cos 2\theta + (\sec^2\theta)\sin 2\theta$
4) $e^{4m}(4\cos 3m - 3\sin 3m)$
5) $3x(1 + 2\log_e x)$
6) $6e^{3t}(3t^2 + 2t - 3)$
7) $1 - 3z + (1 - 6z)\log_e z$
8) a) $\dfrac{1}{(1-x)^2}$ (b) $\dfrac{1 - 2\log_e x}{x^3}$
   (c) $\dfrac{e^x(\sin 2x - 2\cos 2x)}{\sin^2 2x}$
9) $\dfrac{11}{(3 - 4z)^2}$
10) $-\dfrac{2(\sin 2t + \cos 2t)}{e^{2t}}$
11) $-\operatorname{cosec}^2\theta$

Exercise 1.4

1) 21.59 2) -3.77
3) 1.005 4) -0.3573
5) 0 6) 26.15
7) -5.882 8) -4.050
9) 0.5 10) 0.5266

ANSWERS TO CHAPTER 2

Exercise 2.1

1) $18x, 54$ 2) $-187, 254$
3) 29.3 4) $-\dfrac{1}{x^2}, -0.277$
5) $15, 0$ 6) -89.0
7) -0.115 8) 7.52
9) $-4.6, -21.2$
10) 222 11) 2.94

Exercise 2.2

1) 42 m/s   2) -6 m/s^2
3) (a) 6 m/s   (b) 2.41 or -0.41 s
   (c) 6 m/s^2 (d) 1 s

4) -0.074 m/s, 0.074 m/s²
5) 10 m/s, 30 m/s
6) 3.46 m/s
7) (a) 4 rad/s   (b) 36 rad/s²
   (c) 0 s or 1 s
8) (a) -2.97 rad/s
   (b) 0.280 s
   (c) -8.98 rad/s²
   (d) 1.57 s
9) 62.5 kJ

ANSWERS TO CHAPTER 3

Exercise 3.1

1) (a) 11 (max), -16 (min)
   (b) 4 (max), 0 (min)
   (c) 0 (min), 32 (max)
2) (a) 54 (b) $x = 2.5$
   (c) $x = -2$
3) $(3, -15), (-1, 17)$
4) (a) -2 (b) 1
   (c) 9
5) (a) 12 (b) 12.48
6) 15 mm   7) 10 m
8) 4 9) $108\,000$ mm³
10) Radius = Height = 4.57 m
11) 405 mm
12) Diameter = 28.9 mm,
    Height 14.4 mm
13) 5.76 m   14) 2.15
15) 84.8 mm   16) $\cos^2\alpha$

ANSWERS TO CHAPTER 4

Exercise 4.1

1) $\frac{5}{3}x^3 + x^2 + \dfrac{4}{x} + c$

2) $\frac{2}{3}x^{3/2} + 2x^{1/2} + c$

3) $-3\left(\cos\dfrac{x}{3}\right) + c$

4) $\frac{5}{3}(\sin 3\theta) + c$

5) $\phi - \frac{3}{2}(\cos\frac{2}{3}\phi) + c$

6) $2\left(\sin\dfrac{\theta}{2}\right) + \frac{2}{3}\left(\cos\dfrac{3\theta}{2}\right) + c$

7) $t^2 - \frac{1}{2}(\cos 2t) + c$

8) $\frac{1}{3}e^{3x} + c$

9) $-2e^{-0.5u} + c$

10) $\frac{3}{2}e^{2t} - 2e^t + c$

11) $-2e^{-x/2} + \frac{2}{3}e^{3x/2} + c$
12) $(\tan x) + (\log_e x) + c$
13) 7.75 14) 4.67
15) 0.561 16) 0
17) 0.521 18) 0
19) 0.586 20) 1.12
21) 1.33 22) 1.72
23) 0.0585 24) 0.253
25) 4.15 26) 51.1
27) 3.49 28) 2.18
29) 0.811 30) 0.732

Exercise 4.2

1) (a) 0.637 V   (b) 0
2) (a) 3.82 (b) 0
3) 1 volt   4) 0 volt
5) 5 volt   6) 1.67 volt
7) 2.25 volt
8) (a) 0.707 V   (b) 0.707 V
9) (a) 4.24 (b) 4.24
10) $1.42, 2, 5.77, 2.36, 2.45$ volts

ANSWERS TO CHAPTER 5

Exercise 5.1

1) (a) 6.89 (b) 7.12
   (c) 7.00
2) (a) 0.851 (b) 0.846
   (c) 0.848
3) (a) 0.386 (b) 0.387
   (c) 0.386
4) (a) 0.368 (b) 0.365
   (c) 0.366
5) (a) 0.507 (b) 0.495
   (c) 0.499
6) (a) 0.0176 (b) 0.0169
   (c) 0.0171
7) (a) 0.647 (b) 0.647
   (c) 0.647
8) Both 10 and 12 intervals give
   2.42

ANSWERS TO CHAPTER 6

Exercise 6.1

1) $y = \frac{1}{2}x^2 + 1$ 2) $y = x^3 - 128$
3) $y = \frac{1}{3}x^3 - \frac{5}{2}x^2 + k$,
   $y = \frac{1}{3}x^3 - \frac{5}{2}x^2 + 18.7$
4) $y = 1.5x^4 + 1.67x^3 + 7x - 49.4$
5) $y = x^2$

Exercise 6.2

1) $y = Ae^{3x}, y = 2e^{3x}, 807$
2) $s = 12.25e^{-0.560t}, 3.24$
3) $I = 10e^{-33.3t}, 5.14$ A
4) 20 years
5) $Q = 0.0015e^{-41.7t},$
   0.000 652 coulomb
   0.009 72 seconds

ANSWERS TO CHAPTER 7

Exercise 7.1

1) (a) 15.80   (b) 1.968
   (c) 1.094   (d) 0.0314
   (e) 0.581   (f) 0.925
2) (a) 852.8   (b) 39.28
   (c) 2.179
3) 0.003 573   4) 0.011 82
5) 0.5896   6) 25.7
7) 131.8   8) 0.741
9) 44.24   10) 0.005 49
11) 0.769

ANSWERS TO CHAPTER 8

Exercise 8.1

1) $\dfrac{2}{(x-3)} + \dfrac{3}{(x+3)}$

2) $\dfrac{2}{(x-2)} - \dfrac{1}{(x-1)}$

3) $\dfrac{4}{(x+1)} - \dfrac{3}{(2x-1)}$

4) $\dfrac{6}{(x+1)} - \dfrac{5}{(x+2)}$

5) $\dfrac{2}{(x+1)} + \dfrac{3}{(x+2)} + \dfrac{4}{(x+3)}$

6) $\dfrac{1}{(2x-1)} + \dfrac{2}{(x+1)} - \dfrac{1}{(x-1)}$

7) $\dfrac{3}{2x} + \dfrac{1}{(1-x)} - \dfrac{2}{(2-3x)}$

8) $\dfrac{1}{(1+x)} + \dfrac{1}{(x-1)} - \dfrac{4}{(2x+1)}$

9) $\dfrac{1}{(2-x)} + \dfrac{2}{(2-x)^2} - \dfrac{1}{(1+x)}$

10) $\dfrac{1}{(2x+3)} + \dfrac{1}{(x-1)^2} - \dfrac{2}{(x-1)}$

11) $\dfrac{1}{(x+1)} + \dfrac{2}{(x-1)^2}$

12) $\dfrac{2}{(1+x)} - \dfrac{1}{(1+x)^2} - \dfrac{2}{x}$

13) $\dfrac{2x+1}{(x^2+1)} + \dfrac{1}{(x+1)}$

14) $\dfrac{3x+5}{(2+x^2)} - \dfrac{1}{(x-1)}$

15) $\dfrac{1+x}{(2-3x^2)} + \dfrac{3}{(4+x)}$

16) $\dfrac{3x+2}{(x^2+4)} + \dfrac{3}{(x+2)}$

17) $1 + \dfrac{1}{(x-1)} + \dfrac{2}{(x+2)}$

18) $1 + \dfrac{2}{(2x-1)} - \dfrac{1}{(x-1)}$

19) $1 - \dfrac{2}{(x-1)} - \dfrac{1}{(x+1)}$

20) $3 - \dfrac{1}{(x+1)} + \dfrac{1}{(x-2)}$

ANSWERS TO CHAPTER 9

Exercise 9.1

1) $1 + 5z + 10z^2 + 10z^3 + 5z^4 + z^5$
2) $p^6 + 6p^5q + 15p^4q^2 + 20p^3q^3 + 15p^2q^4 + 6pq^5 + q^6$
3) $x^4 - 12x^3y + 54x^2y^2 - 108xy^3 + 81y^4$
4) $32p^5 - 80p^4q + 80p^3q^2 - 40p^2q^3 + 10pq^4 - q^5$
5) $128x^7 + 448x^6y + 672x^5y^2 + 560x^4y^3 + 280x^3y^4 + 84x^2y^5 + 14xy^6 + y^7$
6) $x^3 + 3x + \dfrac{3}{x} + \dfrac{1}{x^3}$
7) $1 + 12x + 66x^2 + 220x^3 + \ldots$
8) $1 - 28x + 364x^2 - 2912x^3 + \ldots$
9) $p^{16} + 16p^{15}q + 120p^{14}q^2 + 560p^{13}q^3 + \ldots$
10) $1 + 30y + 405y^2 + 3240y^3 + \ldots$
11) $x^{18} - 27x^{16}y + 324x^{14}y^2 - 2268x^{12}y^3 + \ldots$
12) $x^{22} + 11x^{18} + 55x^{14} + 165x^{10} + \ldots$

Exercise 9.2

1) $1 + 6x + 15x^2 + 20x^3 + \ldots$
2) $1 + 18x + 144x^2 + 672x^3 + \ldots$
3) $1 - 4x + 10x^2 - 20x^3 + \ldots$
4) $1 + \dfrac{x}{2} + \dfrac{3x^2}{8} + \dfrac{5x^3}{16} + \ldots$
5) $1 - x + x^2 - x^3 + \ldots$
6) $1 - 6x + 24x^2 - 80x^3 + \ldots$
7) $1 + 3x + 6x^2 + 10x^3 + \ldots$
8) $1 + \dfrac{x^2}{3} - \dfrac{x^4}{9} + \dfrac{5}{81}x^6 - \ldots$
9) $1 + \dfrac{3x}{2} + \dfrac{27x^2}{8} + \dfrac{135x^3}{16} + \ldots$
10) $1 - \dfrac{x}{2} - \dfrac{x^2}{24} - \dfrac{5x^3}{432} - \ldots$

Exercise 9.3

2) 1% too large
3) 11% too small
4) 3% decrease
5) 18% increase
6) $\dfrac{3}{\sqrt{2}}\left(1 - \dfrac{x}{4} + \dfrac{3x^2}{32}\right)$
8) 3% decrease
9) 1.5%
11) 2% too small

ANSWERS TO CHAPTER 10

Exercise 10.1

1) (a) 4.48   (b) 22.2
   (c) 0.135   (d) 0.0136
   (e) 1.22   (f) 1.70
   (g) 0.497
2) (a) $1 + 3x + \dfrac{(3x)^2}{2!} + \dfrac{(3x)^3}{3!} + \ldots$

   (b) $1 + 0.5x + \dfrac{(0.5x)^2}{2!} + \dfrac{(0.5x)^3}{3!}$
       $+ \ldots$

   (c) $1 - 1.3x + \dfrac{(1.3x)^2}{2!} - \dfrac{(1.3x)^3}{3!}$
       $+ \ldots$

   (d) $1 - 0.3x + \dfrac{(0.3x)^2}{2!} - \dfrac{(0.3x)^3}{3!}$
       $+ \ldots$
3) (a) 7.051   (b) 1.020
4) $1 + \dfrac{x^2}{2!} + \dfrac{x^4}{4!} + \dfrac{x^6}{6!} + \ldots$

Exercise 10.2

1) 5.47, 0.294
2) $0, -0.0149$
3) (a) $T = 631$ (b) $s = 1.12$
4) 4.34 mA per second
5) (a) 0.18 seconds
   (b) 1200 volts per second
6) (a) 89 800 cells
   (b) 1.35 hours
   (c) 27 900 cells/hour
7) (a) 1.10×10^{-19} grams
   (b) 10^{-10} grams
   (c) 0.578 hours
8) 0.0996

Exercise 10.3

1) $a = 248, b = 34$
2) $\mu = 0.5, k = 5$
3) $C = 2.5, k = 0.01$
4) $c = 76.2, a = 1/15\,000$
5) $A = 40, a = 0.09$
6) $V_0 = 5, a = 0.045$
7) $I = 0.02, T = 0.20$

ANSWERS TO CHAPTER 11

Exercise 11.1

1) (b)   2) (d)   3) (e)
4) (a)   5) (c)   6) (b)
7) (b)   8) (b)   9) (b)
10) $x^2 + y^2 = 100$

Exercise 11.2

1) $a = 70, b = 50$
2) $k_m = 0.016 + \dfrac{0.023}{\mu}$
3) $p = \dfrac{1500}{V}$
4) $m = 4.5, c = 0.5$
5) $k = 0.2$
6) $a = 0.761, b = 10.1, 16.5$ mm
7) $m = 0.040, c = 0.20$
8) $m = 0.1, c = 1.4$
9) $a = 12.5, b = -0.59$

Exercise 11.3

1) $a = 3, n = 2$
2) $a = 2 \times 10^{-6}, n = 4$
3) $n = 2, R = 10$

4) $n = 4$, for $V = 80$ read $V = 70$
5) $t = 0.3 \, \text{m}^{1.5}$
6) $k = 100, n = -1.2$

ANSWERS TO CHAPTER 12

Exercise 12.1

1) (a) 2×2 (b) 2×1
    (c) 3×3 (d) 2×4
2) (a) 9 (b) 4 (c) n^2
3) (a) $\begin{pmatrix} 1 & 3 \\ 2 & 4 \end{pmatrix}$ (b) $(5 \quad -6)$

    (c) $\begin{pmatrix} a & 2 & x \\ b & 3 & -6 \\ 4 & 5 & 0 \end{pmatrix}$ (d) $\begin{pmatrix} 1 & 6 \\ -2 & 2 \\ -3 & 0 \\ -4 & -1 \end{pmatrix}$

4) (a) $\begin{pmatrix} 0 & 0 \\ 9 & 2 \end{pmatrix}$ (b) $\begin{pmatrix} 4 & 2 \\ -3 & 2 \end{pmatrix}$

    (c) $\begin{pmatrix} \frac{5}{6} & \frac{1}{2} \\ \frac{5}{6} & 1 \end{pmatrix}$

5) $a = -8, b = 5, c = 1$

6) $\begin{pmatrix} \frac{1}{3} & \frac{1}{20} \\ \frac{1}{30} & \frac{1}{18} \end{pmatrix}$ **7)** $\begin{pmatrix} 5 & 8 \\ 12 & -2 \end{pmatrix}$

8) $\begin{pmatrix} 0 \\ 5 \end{pmatrix}$

Exercise 12.2

1) (a) $\begin{pmatrix} 6 & 0 \\ -4 & 2 \end{pmatrix}$ (b) $\begin{pmatrix} -12 & 3 \\ 9 & -6 \end{pmatrix}$

    (c) $\begin{pmatrix} -6 & 3 \\ 5 & -4 \end{pmatrix}$ (d) $\begin{pmatrix} 18 & -3 \\ -13 & 8 \end{pmatrix}$

2) (a) $\begin{pmatrix} 14 & 0 \\ 8 & -2 \end{pmatrix}$ (b) $\begin{pmatrix} 2 & 1 \\ 3 & 1 \end{pmatrix}$

    (c) $\begin{pmatrix} 5 & 11 \\ 10 & 22 \end{pmatrix}$ (d) $\begin{pmatrix} a & b \\ c & d \end{pmatrix}$

    (e) $\begin{pmatrix} ka & kb \\ kc & kd \end{pmatrix}$

3) (a) $\begin{pmatrix} 7 & 10 \\ 15 & 22 \end{pmatrix}$ (b) $\begin{pmatrix} 3 & -5 \\ 5 & 8 \end{pmatrix}$

    (c) $\begin{pmatrix} 8 & 10 \\ 20 & 18 \end{pmatrix}$ (d) $\begin{pmatrix} 18 & 15 \\ 40 & 48 \end{pmatrix}$

    (e) $\begin{pmatrix} 13 & 10 \\ 40 & 53 \end{pmatrix}$

Exercise 12.3

1) (a) 24 (b) -6 (c) 14
2) (a) $x = 1, y = 2$
    (b) $x = 4, y = 3$
    (c) $x = 0.5, y = 0.75$

Exercise 12.4

1) $\frac{1}{3}\begin{pmatrix} 4 & -5 \\ -1 & 2 \end{pmatrix}$ **2)** $\begin{pmatrix} 3 & -5 \\ -1 & 2 \end{pmatrix}$

3) $\frac{1}{4}\begin{pmatrix} 2 & -2 \\ -1 & 3 \end{pmatrix}$ **4)** No inverse

5) $\frac{1}{320}\begin{pmatrix} 4 & -24 \\ -24 & 224 \end{pmatrix}$

6) No inverse   **7)** $\frac{1}{7}\begin{pmatrix} 2 & -3 \\ 1 & 2 \end{pmatrix}$

8) $\frac{1}{13}\begin{pmatrix} 5 & 3 \\ -1 & 2 \end{pmatrix}$ **9)** $\begin{pmatrix} 1 & -1 \\ 0 & 1 \end{pmatrix}$

10) (a) $\frac{1}{2}\begin{pmatrix} 2 & 0 \\ -3 & 1 \end{pmatrix}$ (b) $\begin{pmatrix} 2 & -5 \\ -1 & 3 \end{pmatrix}$

    (c) $\frac{1}{2}\begin{pmatrix} 19 & -5 \\ -11 & 3 \end{pmatrix}$

    (d) $\begin{pmatrix} 3 & 5 \\ 11 & 19 \end{pmatrix}$ (e) $\frac{1}{2}\begin{pmatrix} 19 & -5 \\ -11 & 3 \end{pmatrix}$

    (f) equal

Exercise 12.5

1) $x = 6, y = -5$
2) $x = 1, y = 5$
3) $x = 3, y = -1$
4) $x = 2, y = -3$
5) $x = \dfrac{16}{11}, y = \dfrac{9}{11}$
6) $x = 2, y = -5$

ANSWERS TO CHAPTER 13

Exercise 13.1

1) (a) $18 + j10$
    (b) $-j$ (c) $8 - j10$
2) (a) $-1 + j3$
    (b) $-4 - j3$ (c) $10 - j3$

3) (a) $-9 + j21$
   (b) $-36 - j32$
   (c) $-9 + j40$
   (d) 34
   (e) $-21 - j20$
   (f) $18 - j30$
   (g) $0.069 - j0.172$
   (h) $-0.724 + j0.690$
   (i) $-0.138 - j0.655$
   (j) $0.644 + j0.616$
   (k) $3.5 - j0.5$
   (l) $0.2 + j0.6$
4) (a) $1, j0.5$ (b) $0, j5$
   (c) $-3, j3$
5) (a) $-1 \pm j$ (b) $\pm j3$
6) (a) $0.634 - j0.293$
7) (a) $3 + j4$ (b) $0.4 + j0.533$
   (c) $0.692 + j2.538$

Exercise 13.2

2) Mod 5, Arg $53° 8'$;
   Mod 5, Arg $-36°52'$
3) $3.61, 146° 19'$
4) $4.47, -153° 26'$
5) (a) $5 \underline{/36° 52'}$
   (b) $5 \underline{/-53° 8'}$
   (c) $4.24 \underline{/135°}$
   (d) $2.24 \underline{/-153° 26'}$
   (e) $4 \underline{/90°}$ (f) $3.5 \underline{/-90°}$
6) (a) $2.12 + j2.12$
   (b) $-4.49 + j2.19$
   (c) $4.32 - j1.57$
   (d) $-1.60 - j2.77$
7) (a) $56 \underline{/70°}$ (b) $10 \underline{/-50°}$
   (c) $15 \underline{/90°}$ (d) $21 \underline{/-90°}$
8) (a) $2.67 \underline{/-30°}$
   (b) $2 \underline{/-60°}$
   (c) $0.6 \underline{/-9°}$
   (d) $2.83 \underline{/44° 39'}$
9) (a) $30 \underline{/13°}$
   (b) $-2.93 + j6.90$
10) (a) $0.277 \underline{/56° 19'}$
    (b) $13 \underline{/-112° 38'}$
11) (a) $2.83 \underline{/19°}$
    (b) $1.674 + j0.896$
12) (a) $0.172 \underline{/-59° 2'}$
    (b) $0.0427 - j0.0385$

13) (a) $r = 4.5, X_L = 2.2$
    (b) $R = 23, X_C = 35$
    (c) $R = 27.2, X_L = 11.7$
    (d) $R = 6.85, X_C = 1.46$
14) (a) $14° 37'$ (b) 345 watts

ANSWERS TO CHAPTER 14

Exercise 14.1

No answers

Exercise 14.2

4) (a) $\sin\left(\omega t + \dfrac{\pi}{2}\right)$

   (b) $\sin(\omega t - \pi)$
   (c) $\sin \omega t$

   (d) $\sin\left(\omega t - \dfrac{\pi}{6}\right)$

6) $3.61, 33.7°$
7) $v_R = 7.7 \sin(\theta - 19°)$
8) $i_R = 6.6 \sin(\theta + 12°)$
9) $v_R = 51 \sin(\theta - 11°)$

ANSWERS TO CHAPTER 15

Exercise 15.1

1) (a) $-\sin x$ (b) $-\cos x$
   (c) $\cos x$ (d) $\sin x$
   (e) $\tan x$
2) 0.259
4) (a) $\frac{16}{65}$ (b) $\frac{33}{65}$ (c) $\frac{56}{33}$
   (d) $\frac{24}{25}$ (e) $\frac{119}{169}$ (f) $\frac{24}{7}$
6) $46.8°$
7) (b) $0.960, 0.280, -3.429$
8) $10.9°$
9) $18.4°$
10) (a) -0.551 (b) -0.551

Exercise 15.2

1) $3.61 \sin(\theta + 33.7°)$
2) $7.071 \sin(\omega t + 1.429)$
3) $8.602 \sin(\omega t - 0.951)$
4) $8.602, 2.521$ radians
5) $223.6 \sin(300t + 0.464), 223.6$
6) $7.74 \sin(\theta - 18.8°)$
7) $6.57 \sin(\theta + 12.4°)$
8) $51.2 \sin(\theta - 10.6°)$

INDEX